3200

S0-AEE-415

Proceedings in Life Sciences

Axoplasmic Transport in Physiology and Pathology

Edited by D. G. Weiss and A. Gorio

With 76 Figures

Springer-Verlag
Berlin Heidelberg New York 1982

Dr. Dieter G. Weiss

Zoologisches Institut
Universität München
Luisenstraße 14
8000 München 2, FRG

Dr. Alfredo Gorio

Department of Cytopharmacology
Fidia Research Laboratories
Abano Terme, Padova, Italy

QP
363
.A943
1982

ISBN 3-540-11663-X Springer-Verlag Berlin Heidelberg New York
ISBN 0-387-11663-X Springer-Verlag New York Heidelberg Berlin

Library of Congress Cataloging in Publication Data. Main entry under title: Axoplasmic transport in physiology and pathology. (Proceedings in life sciences) Includes index. 1. Axonal transport. 2. Nervous system–Diseases. I. Weiss, D. G. (Dieter), 1945–. II. Gorio, Alfredo. III. Series. [DNLM: 1. Axoplasmic flow. 2. Nervous system–Physiology. 3. Nervous system diseases–Physiopathology. WL 102 A971] QP363.A943 1982   591.1′88   82-10811

This work is subject to copyright. All rights are reserved, whether the whole or part of the material is concerned, specifically those of translation, reprinting, re-use of illustrations, broadcasting, reproduction by photocopying machine or similar means, and storage in data banks. Under § 54 of the German Copyright Law, where copies are made for other than private use, a fee is payable to "Verwertungsgesellschaft Wort", Munich.

© by Springer-Verlag Berlin Heidelberg 1982
Printed in Germany

The use of registered names, trademarks, etc. in this publication does not imply, even in the absence of a specific statement, that such names are exempt from the relevant protective laws and regulations and therefore free for general use.

Offsetprinting and bookbinding: Brühlsche Universitätsdruckerei, Giessen

2131/3130-543210

Preface

Cajal and contemporary scientists have laid the basis of the
modern concepts of the organization of the nervous system: the cir-
cuits of the brain are made up of individual neurons which transfer
information via specialized structures called synapses. Soma and dendrites usually receive the inputs, then the signal is carried all along the
axon to the target areas. To fulfill this task several types of neurons
have developed their unique geometry characterized by a large recep-
tive area (soma and dendrites) and an often very extensive distal
branching with the axon terminals. The volume of cytoplasm which
constitutes the neuronal periphery is often far larger than the cell
body, where the synthetic machinery is located. It is one of the roles
of axoplasmic transport to supply the periphery with proper material
and to sustain the specialized structures necessary for the physiological
activity of the neuron.

Furthermore, it has become more and more clear that target areas
also exert effects on the innervating neurons, and these effects are not
only mediated via recurrent fibers. Synapses have been shown to be
able to pick up material from the synaptic left which is then intra-
axonally transported back to the cell body. This retrograde axoplasmic
transport has therefore been recognized as another basic mechanism
to convey signals from the periphery to the centre. Several articles in
this book deal with the role played by anterograde and retrograde
transport in the physiology of the normal nervous system, in particular
the role played in molecular communication. We now know that in
addition to electrical activity, axoplasmic transport is of crucial importance for the function and development of the nervous system.

General effects of several neurotoxins can be explained by their
specific effects on axoplasmic transport. Furthermore, the pathogenesis of some viral diseases is incomprehensible without the knowledge
of retrograde axoplasmic transport conveying the pathogenic agent
from the periphery to the more central parts of the nervous system.
Many other alterations of nerve function have been found to be correlated with an impairment of axoplasmic transport. Indeed, it is in
most cases not yet clear whether axonal transport causes neuropathies

or whether the disease itself produces changes in axoplasmic transport
as a secondary effect. However, several articles collected in this book
will show a strong link between the occurrence of neuropathies and
alterations of transport.

The publication of this volume was conceived at the occasion of
the *Workshop on Axoplasmic Transport* which has held at Schloss
Elmau in Bavaria in 1981. Written after that meeting, the contributions
to this volume reflect our secured knowledge as well as the questions
still remaining to be answered.

The more basic aspects of axoplasmic transport, i.e., its properties
and our knowledge and current hypotheses about how its molecular
mechanism may work, are the subject of another volume published
concomitantly (D.G. Weiss, ed., *Axoplasmic Transport,* contents see
p. 193). It is revealing that this volume dealing with the physiology and
pathology of axoplasmic transport is the slender one. However, due to
the increasing interest of neurologists, neurosurgeons, neurotoxicol-
ogists, neuropathologists, ophthalmologists, dentists and others dealing
with medical aspects of neurons and the nervous system, this relation is
expected to reverse during the next few years.

The preparation of this volume would not have been possible without the help and understanding of many friends and colleagues to
whom we would like to express our gratitude. We are especially grateful to FIDIA Laboratories, Abano Terme, Italy, for having supported
the preparation of this volume.

München and Abano Terme, August 1982   DIETER G. WEISS
                                              ALFREDO GORIO

Contents

Contents

Contributors

General Properties of Axoplasmic Transport

DIETER G. WEISS [1]

Axoplasmic transport has become a well-established phenomenon whose properties have been described in depth in many publications over the last years. The biological significance of axoplasmic transport is widely recognized. Lacking detailed knowledge of the underlying mechanisms, in this paper all movement of molecules and particles inside nerve cell processes that cannot be related to diffusion or Brownian movement is to be called "axoplasmic transport". This leaves open the question whether one "unitary mechanism" can account for all the phenomena mentioned below, or whether several mechanisms are involved. The term "mechanism" when used in this article is therefore meant to encompass as many different mechanisms as might be needed ultimately to account for all of the phenomena.

This article surveys the general properties of axoplasmic transport which are experimentally verified. It is not intended to cover the literature in detail since an excellent review article was published very recently [41] and since many of these topics are discussed in depth in a recent book [120]. In this article only some, mostly more recent publications are quoted as examples, which do not necessarily represent the first or the only reports on the topics mentioned. If available, preference is given to summarizing and review articles.

The Transported Material

Apparent Lack of Specificity

Axoplasmic transport conveys virtually all axonal and dendritic constituents which have been tested so far. Some of the most common components of nerve cell processes whose transport is well documented are listed in Table 1 without reference to transport velocities and directions. Also, the question whether all or only part of the smaller molecules are bound to macromolecules or organelles during transport is not addressed by Table 1. From this table we can conclude that physicochemical para-

1 Zoologisches Institut, Universität München, Luisenstr. 14, D-8000 München 2, Fed. Rep. Germany

Axoplasmic Transport in Physiology and Pathology
(ed. by D.G. Weiss and A. Gorio)
© Springer-Verlag Berlin Heidelberg 1982

Table 1. Cellular constituents which are axonally transported

Low molecular weight material:	
Amino acids	[16, 45, 64, 100]
Sugars	[8, 29, 121]
Amines	[57]
Transmitters	[17, 102]
Lipids:	[1, 20]
RNA (tRNA):	[56]
Mucopolysaccharides:	[24]
Neurohormones:	[35, 87]
Proteins:	
Enzymes	[11, 12, 15, 17, 111]
Cytoskeletal elements	[83, 119]
Membrane proteins	[71, 78]
Glycoproteins	[24, 40]
Organelles:	
Mitochondria	[32, 65, 101]
Smooth endoplasmic reticulum	[21, 107, 118]
Multivesicular and multilamellate bodies	
(prelysosomal particles)	[107, 118, 126]
Synaptic vesicles	[17, 38]
Neurosecretory granules	[27, 35, 87]
Exogenous material:	
Colloidal gold	[19]
Exogenous proteins	[3, 70, 117]
Horseradish peroxidase	[69, 73, 82]
Lectins	[61, 117]
Tetanus toxin	[26, 96, 115]
Viruses	[67]

meters such as size, charge, shape etc. do not determine whether material is transported or not. Thus, when viewed superficially, one might consider axoplasmic transport to be an unspecific phenomenon. However, it will be discussed below that the dynamics of axoplasmic transport of all these substances and organelles differ considerably and display very specific patterns.

Selectivity

Not all proteins or organelles present in the perikaryon are transported. It seems, however, that the process of selection can be separated from the transport proper. Membrane constituents are thought to be selected when certain proteins and lipids are packaged together into membranous organelles derived from the Golgi apparatus [50, 77] which have been proposed to be the exclusive transport form for rapidly transported material [28, 50]. Similarly, the observed selectivity of the transport in identified neurons for certain transmitters can be explained by selective uptake of these transmitters into storage vesicles rather than by selectivity of the transport process

[38, 103]. Ribosomes and dictyosomes are apparently prevented by unknown mechanisms from passing the axon hillock. Again, such mechanisms must be considered not to be part of the transport mechanism since various kinds of exogenous materials, once in the axon, are transported very nicely (Table 1).

The process of loading the material onto the transport machinery may also give rise to the "routing" phenomenon, i.e., transport of various materials in different proportions along the central and peripheral branches of axons of, e.g., dorsal root ganglion cells [83, 92].

Dynamics of Transport

Transport is Directed

Transport may be anterograde (orthograde), that is leaving the perikaryon, or retrograde, that is directed towards the perikaryon. Transport in both directions occurs also in dendrites [79]. The anterogradely and retrogradely transported organelles are different from each other [107, 118]. Reversal of direction of optically detectable organelles is a rare event in undisturbed axons [30, 107].

Behavior at Obstructions

Endogenous or exogenous obstructions of the axon such as nodes of Ranvier, terminals, ligatures, crush or cold blocks cause accumulations of anterogradely and retrogradely transported material at their respective sides [5, 17, 106, 107, 118, 126]. In these cases there is initially no general damming of the axoplasm, but the rapidly transported organelles move towards the obstruction being stopped only in its immidiate vicinity. Part of the material is reversed if the block persists for some time ("turnaround") [6, 107, 110].

Saltatory Movement

Optically detectable organelles (mitochondria, prelysosomal particles) move in a saltatory fashion with periods of movement and periods of rest [31, 97, 109]. The average transport velocity of a population of organelles (see Table 2) is therefore considerably less than the actual saltation velocity of $1-3$ $\mu m/s$. The dynamics of the saltations are very similar for both directions [108]. The saltation velocities of the visible organelles do not vary with particle size (c.f. [123]). It is unknown whether small organelles below the resolution of the microscope or soluble molecules behave similarly.

Table 2. Characteristic population velocities of materials transported in optic nerve

Transport group	Maximal velocity mm/d	Material transported	Identified components	Destination
I	240–410	Membranous material (tubulo-vesicular organelles), soluble material	Na^+, K^+-ATPase, GAP's	Axon and axon terminal
	20–240	Synaptic vesicles	Protein I (synaptin I)	Axon terminal
II	20–70	Mitochondria and possibly other membrane bounded organelles	F_1 ATPase, fodrin	Axon and axon terminal
III–IV (SCb)	2–20	Axoplasmic matrix ("microtrabeculae"), subaxolemmal specializations, mostly "soluble" proteins	Myosin-like proteins M1 and M2, actin, calmodulin, clathrin, enolase, fodrin, creatine phosphokinase	Axon
V (SCa)	0.5–2	Cytoskeletal elements	α- and β-tubulin, fodrin, 3 neurofilament proteins	axon

Modified from [2] and [72]; for references see [2, 41, 72, 123].
Abbreviations: SCa, SCb: slow components a and b according to Lasek's nomenclature [72, 119]

Constance of Velocity

If the velocity of a whole population of radioactively labeled compounds is determined, one observes that the maximal transport velocities are constant all along the axons [13, 44, 88]. This also holds true for low molecular weight substances [121, 124] whose movement has often been ascribed to diffusion [60, 81]. Constant transport velocities, however, cannot be due to diffusion [100].

Radioactivity Profiles

Characteristic radioactivity profiles occur along the axon if transport is demonstrated in pulse-chase experiments, i.e., if radioactive precursor is available only on one side of the axon and for a limited period of time. The characteristic properties of these profiles are: a leading wave or peak is followed by a plateau or saddle region and the peak is asymmetric, reveals tailing and broadens with time [37, 44, 86, 89, 90]. In rapid transport even the broadening of the peak cannot be explained by diffusion alone [46]. The amount of radioactivity in the moving peaks decreases with time and distance [44, 86, 121]. Similar profiles are obtained for slowly transported materials including cytoskeletal elements in which case the peak broadening is slower [13, 55] and occurs with the velocity expected from diffusion of free protein [46]. Similar radioactivity profiles have been shown also for low molecular weight substances such as amino acids and sugars, although it is not known whether these are transported as free molecules or sequestered in storage organelles [121, 124].

Transport Velocities

Various types of organelles and molecules move with different, but specific average velocities ranging from 0.5 mm/day up to 370 mm/day. The maximal velocity is in many vertebrate and mollusc nerves 410 mm/day (at 37°C) [23, 44, 88, 104, 121]. Table 2 gives an overview of the transport velocities of a variety of cell constituents. Within the broad spectrum of velocities there are five regions which have been found in mammalian optic nerves to be especially well populated with specific polypeptides [55, 62, 125]. There are, however, substances known to move with intermediate velocities or with a wider range of velocities so that they are reported to be conveyed with more than one transport group (e.g. fodrin, Table 2). Transport groups I–II contain heterogeneous, mostly membrane-bound material and their velocities seem to be very similar for most vertebrates (if temperature is corrected for) [2, 88].

The velocities of the more slowly transported materials vary much more between different species and different nerves (groups IV and V). Apparently the organization of the cytoskeleton is of importance since in nerves with fewer neurofilaments such as the vagus and olfactory nerves the slowest components move with 8–25 mm/day whereas in the optic nerves 0.5–4 mm/day are observed [7, 13, 33, 116]. Table 2 gives the values obtained for optic nerves, however many more data on slow transport in other nerves are required, before we can make more general statements on slow transport velocities.

The question how these components are organized while undergoing transport is presently being debated. The idea was proposed that each one of the five transport groups represents one structural component of the cell [119]. This does not seem to fit the data on the rapid transport groups which contain very heterogenous material [2, 121, 123]. For the slower groups this idea seems to be persuasive in the case of the findings on optic nerves [72], whereas the data on other nerves are less consistent (as is also discussed in [123]).

Retrogradely transported material has been reported to move with rapid velocities of at least 100–290 mm/day [12, 105, 117], but recently also a slow component (3–6 mm/day) has been found [34]. The retrogradely moving material consists mainly of multivesiculate and multilamellar organelles which seem to be destined for degradation in the perikaryon [107, 118]. This material originates from endocytosis at the synaptic level where synaptic membrane components as well as extracellular material are sampled [68, 117] and from the reversal of axonal constituents which are also conveyed anterogradely but in part reverted [3, 4].

Particle Size and Transport Velocity

There seems to exist a weak biphasic correlation of population transport velocity and particle size [47]. The most rapidly transported material (group I) consists of smooth-walled tubulo-vesicular structures of about 50 nm in diameter and 180 nm in length [107, 118]. Larger material (mitochondria, neurosecretory granules and retrogradely moving prelysosomal particles) seem to move at slower population velocities (group II; [12, 78, 87, 105, 107, 115]). Smaller material seems also to be considerably retarded, although part of it is initially transported rapidly [47, 100]. The subcellular status of such rapidly transported soluble proteins and low molecular weight compounds is an unclarified issue [28, 61, 121, 123]. One could conclude that there is an optimal particle size in the range of maximally 80 nm of diameter, whereas smaller and larger materials move slower. Since the velocity of individual particle saltations is not size dependent [65, 123], the pauses between the saltations must increase for larger organelles.

This correlation seems not to apply to the movement of the group of slowly transported soluble proteins (Group IV) and the cytoskeletal constituents (group V).

Temperature Dependence

The velocities of rapidly and slowly transported materials are within physiological limits linearly [13, 42] or nearly linearly temperature dependent [10, 23]. The maximal transport velocities seem to be only little influenced by length, diameter or functional type of the axon, by the animal species (except arthropods) or by the electrical activity of the axon [41].

Properties of the Transport Mechanism

Independence from the Perikaryon

Removal of the perikaryon or dissection of axonal segments does not impair transport for several hours if the nerve is kept under physiological conditions [44, 51]. This leads to the conclusion that the mechanism, at least for rapid transport, is inherent in any segment of the axon.

ATP, Ca^{2+}, Microtubules, Actin

The mechanism of axoplasmic transport requires ATP [75], microtubules (pharmacological evidence: [52, 66]; morphological evidence: [123]; theoretical evidence: [48]) and most probably also actin (pharmacological evidence: [36, 58]). Ca^{2+} is required for the synthesis and formation of the rapidly transported organelles at the level of the Golgi apparatus [49, 77] and most probably also during transport in the axon ([50, 59, 93], but see also [28, 108]).

Guiding Elements and Directionality

It is generally believed that similarly to other forms of cellular motility axoplasmic transport also depends on the presence of guiding structures (e.g., [84, 104]). This can also be postulated on theoretical grounds since the transport direction can most easily be defined by a polar structure [122, 123]. For both purposes only cytoskeletal elements are suitable. Out of these, the neurofilaments seem to be located in the axon where rapid transport does not take place [32, 39, 94], actin filaments are only randomly oriented in axons [39, 74], whereas microtubules could well fulfill such functions (cf. [122]). Autoradiographic work supports the view that rapid transport takes place in axonal regions where microtubules, mitochondria and smooth endoplasmic reticulum are located [21, 94].

Force-Generating Mechanism

The molecular mechanism which links ATP consumption to the generation of the shear force is unknown. It is generally believed that this is accomplished by force-generating enzymes (ATPases). If such enzymes are involved, they would work effectively only if they were oriented in the proper direction and attached to some structure massive enough to absorb the recoil (for review see [122]). Both dynein-like [28] and actomyosin-like ATPases (e.g., [63]) were proposed, but other, not yet characterized ATPases are also visualizable.

Since a whole variety of cellular constituents are to be transported, these could either be confined in one kind of general transport organelle upon which the shear force then could act specifically, or, if shear force is acting in a non-specific manner upon all the axoplasmic components, no such transport organelle would be needed [43, 122].

Energy Consumption

The energy available for axoplasmic transport can be estimated to be of the order of 5×10^{-9} erg s^{-1} for 1 μm^3 of axoplasm [48]. Since the viscosity of axoplasm is very high for moving organelles [99] organelle movement can for energetical reasons only take place in low viscosity channels between cytoskeletal elements or after local destruction of such elements — provided that the laws of fluid dynamics can be applied also to cytoplasm [48, 98].

Reversible Interactions

There is good evidence that the transported material and the mechanism interact reversibly and intermittently. This is conspicuous in the case of visible material as saltatory movement [108] but it can also be concluded more generally from effects such as axonal retention, tailing, deposition and cooperativity [44, 80, 86, 90, 104]. Such behavior may lead to a spectrum of velocities [85, 123], and it can formally be described analogous to chromatography [113].

Different Transport Velocities

The transport velocity differences can be explained by several means, none of which can be excluded from our present knowledge. (1) Several different mechanisms, each having a characteristic velocity, convey different materials (e.g., [119]). (2) One rapid transport mechanism conveys all materials, the different velocities being due to different duration or probability of these materials to stay on the mechanism (chromatography) (e.g., [14, 47, 90]). (3) One rapid mechanism conveys activily only the rapid material, while other material is moved by drag according to its physical properties (e.g., [76]).

Directionality

The dynamics of anterograde and retrograde transport are very similar (e.g., [108]) so that one would assume the same kind of mechanism to underly both. How shear forces can be generated in the two directions differing by $180°$ is unknown.

Regulation of Axoplasmic Transport

Even in extreme physiological states of the neuron such as in situations of blocked or increased electrical and exocytotic activity at the synapse there seems to be no significant influence on axoplasmic transport. This is especially true for the transport velocity which virtually cannot be influenced by altering the physiological activity of the neuron (see [41], p. 1188, for review). The amount of transported material can be increased in certain situations of physiological stimulation, especially in the hypothalamo-neurohypophysial system [41, 87]. This may be explained, however, by increased perikaryal synthesis of the transported material, since it is known that transport has a considerable excess capacity and can in some cases accommodate four times the amount of material transported in the resting state [9]. That axoplasmic transport cannot be regulated very effectively is also evident from the fact that by electrical stimulation synapses can be depleted of vesicles and transmission be abolished [54, 127]. Elevated velocities could be observed only if the amount of organelles was increased [37, 103]. Davison [18] had already speculated that due to the lack of regulation the neuron has always to convey a surplus of material, thus necessitating retrograde transport (which itself seems not be regulated either [114]).

There are many proposed models of how axonal transport can be regulated intracellularly [22, 25, 32, 53, 91, 95, 112]. Whether the neuron makes use of any of these mechanisms remains to be elucidated by further experimentation.

Conclusion

The above description of properties of axoplasmic transport could serve as a substitute for a more precise definition of what we understand by "axoplasmic transport". The properties have been discussed on three levels: (1) The description of the transported material. Our knowledge of this is good except that we have to determine more precisely the state of organization these materials have while undergoing transport. (2) The description of transport dynamics. Here more detailed information is required especially on slower velocity groups, on the turnaround process and on retrograde transport before we can draw definitive conclusions on the underlying mechanism. (3) The properties of the transport mechanism(s) are least well known and as our knowledge is poor this field is wide open to hypotheses which are very numerous (see [120] for some and [41, 122] for review).

Acknowledgment. I am grateful to Sepp Gulden, without whom this article would not have been written.

References

1. Abe T, Haga T, Kurokawa M (1973) Rapid transport of phosphatidylcholine occuring simultaneously with protein transport in the frog sciatic nerve. Biochem J 136:731–740
2. Baitinger C, Levine J, Lorenz T, Simon C, Skene P, Willard M (1982) Characteristics of axonally transported proteins. In: Weiss DG (ed) Axoplasmic transport. Springer, Berlin Heidelberg New York, pp 110–120
3. Bisby MA (1980) Retrograde axonal transport. In: Fedoroff et al. (eds) Advances in cellular neurobiology, vol I. Acad Press, New York, pp 69–117
4. Bisby MA (1982) Retrograde axonal transport of endogenous proteins. In: Weiss DG (ed) Axoplasmic transport. Springer, Berlin Heidelberg New York, pp 193–199
5. Bisby MA (1982) Ligature techniques. In: Weiss DG (ed) Axoplasmic transport. Springer, Berlin Heidelberg New York, pp 437–441
6. Bisby MA, Bulger VT (1977) Reversal of axonal transport at a nerve crush. J Neurochem 29:313–320
7. Brady ST, Lasek RJ (1982) The slow components of axonal transport: movements, composition and organization. In: Weiss DG (ed) Axoplasmic transport. Springer, Berlin Heidelberg New York, pp 206–217
8. Breer H, Rahmann H (1974) Axonal transport of [^{3}H]glucose radioactivity in the optic system of *Scardinius erythrophthalmus.* J Neurochem 22:245–250
9. Brimijoin S (1979) On the kinetics and maximal capacity of the system for rapid axonal transport in mammalian neurons. J Physiol 292:325–337
10. Brimijoin S, Olsen J, Rosenson R (1979) Comparison of the temperature-dependence of rapid axonal transport and microtubules in nerves of the rabbit and bullfrog. J Physiol 287:303–314
11. Brimijoin S, Wiermaa MJ (1977) Rapid axonal transport of tyrosine hydroxylase in rabbit sciatic nerves. Brain Res 121:77–96
12. Brimijoin S, Wiermaa MJ (1978) Rapid orthograde and retrograde axonal transport of acetylcholinesterase as characterized by the stop-flow technique. J Physiol 285:129–142

13. Cancalon P (1979) Influence of temperature on the velocity and on the isotope profile of slowly transported labeled proteins. J Neurochem 32:997–1007

14. Copeland AR (1976) Axonal transport – II. Convection. Bull Math Biol 38:435–444

15. Couraud J-Y, Di Giamberardino L (1982) Axonal transport of the molecular forms of acetylcholinesterase. Its reversal at a nerve transection. In: Weiss DG (ed) Axoplasmic transport. Springer, Berlin Heidelberg New York, pp 144–152

16. Csanyi V, Gervai J, Lajtha A (1973) Axoplasmic transport of free amino acids. Brain Res 56:271–284

17. Dahlström A (1971) Axoplasmic transport (with particular respect to adrenergic neurons). Phil Trans R Soc London Ser B 261:325–358

18. Davison PF (1970) Axoplasmic transport: physical and chemical aspects. In: Schmitt FO (ed) The neurosciences: Second study program. Rockefeller Univ Press, New York, pp 851–857

19. De Lorenzo AJD (1970) The olfactory neuron and the blood-brain barrier. In: Wolstenholme J, Knight J (eds) Taste and smell in vertebrates. Churchill, London

20. Droz B, Brunetti M, Di Giamberardino L, Koenig HL, Porcellati G (1982) Axoplasmic transport and axon-glia transfer of phospholipid constituents. In: Weiss DG (ed) Axoplasmic transport. Springer, Berlin Heidelberg New York, pp 170–174

21. Droz B, Rambourg A, Koenig HL (1975) The smooth endoplasmic reticulum: structure and role in the renewal of axonal membrane and synaptic vesicles by fast axonal transport. Brain Res 93:1–13

22. Durham ACH (1974) A unified theory of the control of actin and myosin in nonmucle movements. Cell 2:123–136

23. Edström A. Hanson M (1973) Temperature effects on fast axonal transport of proteins in vitro in frog sciatic nerves. Brain Res 58:345–354

24. Elam JS, Peterson NW (1976) Axonal transport of sulfated glycoproteins and mucopolysaccharides in the garfish olfactory nerve. J Neurochem 26:845–850

25. Ellisman MH (1982) A hypothesis for rapid axoplasmic transport based upon focal interactions between axonal membrane systems and the microtrabecular crossbridges of the axoplasmic matrix. In: Weiss DG (ed) Axoplasmic transport. Springer, Berlin Heidelberg New York, pp 390–396

26. Erdmann G, Wiegand H, Wellhöner HH (1975) Intraaxonal and extraaxonal transport of ^{125}I-tetanus toxin in early local tetanus. Naunyn-Schmiedeberg's Arch Pharmacol 290:357–373

27. Flament-Durand J, Couck A-M, Dustin P (1975) Studies on the transport of secretory granules in the magnocellular hypothalamic neurons of the rat. II. Action of vincristine on axonal flow and neurotubules in the paraventricular and supraoptic nuclei. Cell Tissue Res 164:1–9

28. Forman DS (1982) Saltatory organelle movement and the mechanism of fast axonal transport. In: Weiss DG (ed) Axoplasmic transport. Springer, Berlin Heidelberg New York, pp 234–240

29. Forman DS, McEwen BS, Grafstein B (1971) Rapid transport of radioactivity in goldfish optic nerve following injections of labeled glucosamine. Brain Res 28:119–130

30. Forman DS, Padjen AL, Siggins GR (1975) Movement of organelles in living nerve fibers (scientific film). National Audio Visual Center, Washington

31. Forman DS, Padjen AL, Siggins GR (1977) Axonal transport of organelles visualized by light microscopy: cinemicrographic and computer analysis. Brain Res 136:197–213

32. Friede RL, Ho K-C (1977) The relation of axonal transport of mitochondria with microtubules and other axoplasmic organelles. J Physiol 265:507–519

33. Frizell M, Sjöstrand J (1974) The axonal transport of slowly migrating [3H]-leucine labeled proteins and regeneration rate in regenerating hypoglossal and vagus nerves of the rabbit. Brain Res 81:267–283

34. Gainer H, Fink DJ (1982) Covalent labelling techniques and axonal transport. In: Weiss DG (ed) Axoplasmic transport. Springer, Berlin Heidelberg New York, pp 464–470

35. Gainer H, Sarne Y, Brownstein MJ (1977) Neurophysin biosynthesis: conversion of a putative precursor during axonal transport. Science 195:1354–1356

36. Goldberg DJ, Harris DA, Lubit BW, Schwartz JH (1980) Analysis of the mechanism of fast axonal transport by intracellular injection of potentially inhibitory macromolecules: Evidence for a possible role of actin filaments. Proc Natl Acad Sci USA 77:7448–7452

37. Goldberg DJ, Schwartz JH, Sherbany AA (1978) Kinetic properties of normal and perturbed axonal transport of serotonin in a single identified axon. J Physiol 281:559–579

38. Goldman JE, Schwartz JH (1974) Cellular specificity of serotonin storage and axonal transport in identified neurones of *Aplysia californica*. J Physiol 242:61–76

39. Goldman RD, Chojnacki B, Goldman AE, Starger J, Steinert P, Talian J, Whitman M, Zackroff R (1981) Aspects of the cytoskeleton and cytomusculature of nonmucle cells. Neurosci Res Prog Bull 19:59–82

40. Goodrum JF, Toews AD, Morell P (1979) Axonal transport and metabolism of [^{3}H]fucose and [^{35}S]sulfate-labeled macromolecules in the rat visual system. Brain Res 176:255–272

41. Grafstein B, Forman DS (1980) Intracellular transport in neurons. Physiol Rev 60:1167–1283

42. Gross GW (1973) The effect of temperature on the rapid axoplasmic transport in C-fibers. Brain Res 56:359–363

43. Gross GW (1975) The microstream concept of axoplasmic and dendritic transport. Adv Neurol 12:283–296

44. Gross GW, Beidler LM (1975) A quantitative analysis of isotope concentration profiles and rapid transport velocities in the C-fibers of the garfish olfactory nerve. J Neurobiol 6:213–232

45. Gross GW, Kreutzberg GW (1978) Rapid axoplasmic transport in the olfactory nerve of the pike: I. Basic transport parameters for proteins and amino acids. Brain Res 139:65–76

46. Gross GW, Stewart GH, Horwitz B (1981) Molecular diffusion cannot account for spreading of isotope distribution peaks during rapid axoplasmic transport. Brain Res 216:215–218

47. Gross GW, Weiss DG (1977) Subcellular fractionation of rapidly transported axonal material in olfactory nerve: evidence for a size-dependent molecule separation during transport. Neurosci Lett 5:15–20

48. Gross GW, Weiss DG (1982) Theoretical considerations on rapid transport in low viscosity axonal regions. In: Weiss DG (ed) Axoplasmic transport. Springer, Berlin Heidelberg New York, pp 330–341

49. Hammerschlag R (1982) Multiple roles of calcium in the initiation of fast axonal transport. In: Weiss DG (ed) Axoplasmic transport. Springer, Berlin Heidelberg New York, pp 279–286

50. Hammerschlag R, Stone GC (1982) Fast axonal transport as endomembrane flow. In: Weiss DG (ed) Axoplasmic transport. Springer, Berlin Heidelberg New York, pp 406–413

51. Hanson M, Bergqvist JE (1982) In vitro chamber systems to study axonal transport. In: Weiss DG (ed) Axoplasmic transport. Springer, Berlin Heidelberg New York, pp 429–436

52. Hanson M, Edström A (1978) Mitosis inhibitors and axonal transport. Int Rev Cytol Suppl 7:373–402

53. Henkart MP, Reese TS, Brinley FJ (1978) Endoplasmic reticulum sequesters calcium in the squid giant axon. Science 202:1300–1303

54. Heuser JE, Reese TS (1973) Evidence for recycling of synaptic vesicle membrane during transmitter release at the frog neuromuscular junction. J Cell Biol 57:315–344

55. Hoffman PN, Lasek RJ (1975) The slow component of axonal transport. Identification of major structural polypeptides of the axon and their generality among mammalian neurons. J Cell Biol 66:351–366

56. Ingoglia NA, Zanakis MF (1982) Axonal transport of 4S RNA. In: Weiss DG (ed) Axoplasmic transport. Springer, Berlin Heidelberg New York, pp 161–169

57. Ingoglia NA, Sturman JA, Eisner RA (1977) Axonal transport of putrescine, spermidine and spermine in normal and regenerating goldfish optic nerves. Brain Res 130:433–445

58. Isenberg G, Schubert P, Kreutzberg GW (1980) Experimental approach to test the role of actin in axonal transport. Brain Res 194:588–593

59. Kanje M, Edström A, Ekström P (1982) The role of Ca^{2+} in rapid axonal transport. In: Weiss DG (ed) Axoplasmic transport. Springer, Berlin Heidelberg New York, pp 294–300

60. Karlsson J-O (1977) Is there an axonal transport of amino acids? J Neurochem 29:615–617

61. Karlsson J-O (1982) Isolation and characterization of undenaturated rapidly transported proteins. In: Weiss DG (ed) Axoplasmic transport. Springer, Berlin Heidelberg New York, pp 121–124
62. Karlsson J-O, Sjöstrand J (1971) Synthesis, migration and turnover of protein in retinal ganglion cells. J Neurochem 18:749–767
63. Kerkut GA (1975) Axoplasmic transport. Comp Biochem Physiol A 51:701–704
64. Kidwai AM, Ochs S (1969) Components of fast and slow phases of axoplasmic flow. J Neurochem 16:1105–1112
65. Kirkpatrick JB, Stern LZ (1973) Axoplasmic flow in human sural nerve. Arch Neurol 28: 308–312
66. Kreutzberg GW (1969) Neuronal dynamics and axonal flow. IV. Blockade of intra-axonal enzyme transport by colchicine. Proc Natl Acad Sci USA 62:722–728
67. Kristensson K (1970) Transport of fluorescent protein tracer in peripheral nerves. Acta Neuropathol 16:293–300
68. Kristensson K (1982) Retrograde axonal transport of exogenous macromolecules. In: Weiss DG (ed) Axoplasmic transport. Springer, Berlin Heidelberg New York, pp 200–205
69. Kristensson K , Olsson Y (1971) Uptake and retrograde axonal transport of peroxidase in hypoglossal neurons. Electron microscopical localization in neuronal perikaryon. Acta Neuropathol 19:1–9
70. Kristensson K, Olsson Y, Sjöstrand J (1971) Axonal uptake and retrograde transport of exogenous proteins in the hypoglossal nerve. Brain Res 32:399–406
71. Krygier-Brévart V, Weiss DG, Mehl E, Schubert P, Kreutzberg GW (1974) Maintenance of synaptic membranes by the fast axonal flow. Brain Res 77:97–110
72. Lasek RJ, Brady ST (1982) The structural hypothesis of axonal transport: two classes of moving elements. In: Weiss DG (ed) Axoplasmic transport. Springer, Berlin Heidelberg New York, pp 397–405
73. LaVail JH, LaVail MM (1974) The retrograde intraaxonal transport of horseradish peroxidase in the chick visual system: a light and electron microscopic study. J Comp Neurol 157:303–358
74. LeBeux XJ, Willemot J (1975) An ultrastructural study of the microfilaments in rat brain by means of heavy meromyosin labeling. I. The perikaryon, dendrites and the axon. Cell Tiss Res 160:1–36
75. Leone J, Ochs S (1978) Anoxic block and recovery of axoplasmic transport and electrical excitability of nerve. J Neurochem 9:229–245
76. Levine J, Simon C, Willard M (1982) Mechanistic implications of the behavior of axonally transported proteins. In: Weiss DG (ed) Axoplasmic transport. Springer, Berlin Heidelberg New York, pp 275–278
77. Longo FM, Hammerschlag R (1980) Relation of somal lipid synthesis to the fast axonal transport of protein and lipid. Brain Res 193:471–485
78. Lorenz T, Willard M (1978) Subcellular fractionation of intraaxonally transported polypeptides in the rabbit visual system. Proc Natl Acad Sci USA 75:505–509
79. Lynch G, Smith RL, Browning MD, Deadwyler S (1975) Evidence for bidirectional dendritic transport of horseradish peroxidase. Adv Neurol 12:297–311
80. Mackey S, Schuessler G, Goldberg DJ, Schwartz JH (1981) Dependence of fast axonal transport on the local concentration of organelles. Biophys J 36:455–459
81. Magid A (1973) Axonal transport – simple diffusion? Science 182:180
82. Mesulam M-M (1982) Tracing neural connections with horseradish peroxidase. John Wiley, Chichester
83. Mori H, Komiya Y, Kurokawa M (1979) Slowly migrating axonal polypeptides. Inequalities in their rate and amount of transport between two branches of bifurcating axons. J Cell Biol 82:174–184
84. Morré DJ (1982) Intracellular vesicular transport: vehicles, guide elements, mechanisms. In: Weiss DG (ed) Axoplasmic transport. Springer, Berlin Heidelberg New York, pp 2–14
85. Muñoz-Martínez EJ (1982) Axonal retention of transported material and the lability of nerve terminals. In: Weiss DG (ed) Axoplasmic transport. Springer, Berlin Heidelberg New York, pp 267–274
86. Muñoz-Martínez EJ, Núñez R, Sanderson A (1981) Axonal transport: a quantitative study of retained and transported protein fraction in the cat. J Neurobiol 12:15–26

87. Norström A (1975) Axonal transport and turnover of neurohypophysial proteins in the rat. Ann NY Acad Sci 248:46–63
88. Ochs S (1972) Rate of fast axoplasmic transport in mammalian nerve fibers. J Physiol 227: 627–645
89. Ochs S (1972) Fast transport of materials in mammalian nerve fibers. Science 176:252–260
90. Ochs S (1975) Retention and redistribution of proteins in mammalian nerve fibres by axoplasmic transport. J Physiol 253:459–475
91. Ochs S (1982) On the mechanism of axoplasmic transport. In: Weiss DG (ed) Axoplasmic transport. Springer, Berlin Heidelberg New York, pp 342–350
92. Ochs S, Erdman J, Jersild RA, McAdoo V (1978) Routing of transported materials in the dorsal root and nerve fiber branches of the dorsal root ganglion. J Neurobiol 9:465–481
93. Ochs S, Worth RM, Chan S-Y (1977) Calcium requirement for axoplasmic transport in mammalian nerve. Nature (London) 270:748–750
94. Papasozomenos SCh, Autilio-Gambetti L, Gambetti P (1982) The IDPN axon: Rearrangement of axonal cytoskeleton and organelles following β,β'-iminodipropionitrile (IDPN) intoxication. In: Weiss DG (ed) Axoplasmic transport. Springer, Berlin Heidelberg New York, pp 241–250
95. Porter KR, Byers HR, Ellisman MH (1979) The cytoskeleton. In: Schmitt FO, Worden FG (eds) The neurosciences, 4th study program. MIT Press, Cambridge, pp 703–722
96. Price DL, Griffin J, Young A, Peck K, Stocks A (1975) Tetanus toxin: direct evidence for retrograde intraaxonal transport. Science 188:945–947
97. Rebhun LI (1972) Polarized intracellular particle transport: saltatory movements and cytoplasmic streaming. Int Rev Cytol 32:93–137
98. Rebhun LI, Sander G (1971) Electron microscope studies of frozen-substituted marine eggs. III. Structure of the mitotic apparatus of the first meiotic division. Am J Anat 130: 35–54
99. Rubinson KA, Baker PF (1979) The flow properties of axoplasm in a defined chemical environment: influence of anions and calcium. Proc R Soc London B 205:323–345
100. Schmid G, Wagner L, Weiss DG (1982) Rapid axoplasmic transport of free leucine. J Neurobiol, in press
101. Schmidt RE, McDougal DB (1978) Axonal transport of selected particle-specific enzymes in rat sciatic nerve in vivo and its response to injury. J Neurochem 30:527–535
102. Schwartz JH (1979) Axonal transport: components, mechanisms, and specificity. Ann Rev Neurosci 2:467–504
103. Schwartz JH, Goldberg DJ (1982) Studies on the mechanism of fast axoplasmic transport in single identified neurons. In: Weiss DG (ed) Axoplasmic transport. Springer, Berlin Heidelberg New York, pp 351–361
104. Schwartz JH, Goldman JE, Ambron RT, Goldberg DJ (1975) Axonal transport of vesicles carrying serotonin in the metacerebral neuron of *Aplysia californica.* Cold Spring Harbor Symp Quant Biol 40:83–92
105. Shield LK, Griffin JW, Drachman DB, Price DL (1977) Retrograde axonal transport: a direct method for measurement of rate. Neurology 27:393
106. Smith DS, Järlfors U, Beránek R (1970) The organization of synaptic axoplasm in the lamprey *(Petromyzon marinus)* central nervous system. J Cell Biol 46:199–219
107. Smith RS (1980) The short term accumulation of axonally transported organelles in the region of localized lesions of single myelinated axons. J Neurocytol 9:39–65
108. Smith RS (1982) Axonal transport of optically detectable particulate organelles. In: Weiss DG (ed) Axoplasmic transport. Springer, Berlin Heidelberg New York, pp 181–192
109. Smith RS, Koles ZJ (1976) Mean velocity of optically detected intra-axonal particles measured by a cross-correlation method. Can J Physiol Pharmacol 54:859–869
110. Snyder RE, Smith RS (1982) Application of position-sensitive detectors to the study of the axonal transport of β-emitting isotopes. In: Weiss DG (ed) Axoplasmic transport. Springer, Berlin Heidelberg New York, pp 442–453
111. Starkey RR, Brimijoin S (1979) Stop-flow analysis of the axonal transport of DOPA decarboxylase (EC 4.1.1.26) in rabbit sciatic nerves. J Neurochem 32:437–441

112. Stearns ME (1980) Lattice-dependent regulations of axonal transport by calcium ions. In: De Brabander M, De Mey J (eds) Microtubules and microtubule inhibitors 1980. Elsevier/ North Holland, Amsterdam, pp 17–30

113. Stewart GH, Horwitz B, Gross GW (1982) A chromatographic model of axoplasmic transport. In: Weiss DG (ed) Axoplasmic transport. Springer, Berlin Heidelberg New York, pp 414–422

114. Stöckel K, Dumas M, Thoenen H (1978) Uptake and subsequent retrograde axonal transport of nerve growth factor (NGF) are not influenced by neuronal activity. Neurosci Lett 10:61–64

115. Stöckel K, Schwab M, Thoenen H (1975) Comparison between the retrograde axonal transport of nerve growth factor and tetanus toxin in motor, sensory and adrenergic neurons. Brain Res 99:1–16

116. Tashiro T, Kurokawa M (1982) Rapid transport of a calmodulin-related polypeptide in the vagal nerve. In: Weiss DG (ed) Axoplasmic transport. Springer, Berlin Heidelberg New York, pp 153–160

117. Thoenen H, Schwab M (1978) Physiological and pathophysiological implications of retrograde transport of macromolecules. In: Adv Pharmacol Therap 5, Neuropsychopharmacology. Pergamon Press, Oxford New York, pp 37–59

118. Tsukita S, Ishikawa H (1980) The movement of membraneous organelles in axons. Electron microscopic identification of anterogradely and retrogradely transported organelles. J Cell Biol 84:513–530

119. Tytell M, Black MM, Garner JA, Lasek RJ (1981) Axonal transport: each major rate component reflects the movement of distinct macromolecular complexes. Science 214:179–181

120. Weiss DG (ed) (1982) Axoplasmic transport. Springer, Berlin Heidelberg New York, 470 p

121. Weiss DG (1982) 3-O-methyl-D-glucose and β-alanine: Rapid axoplasmic transport of metabolically inert low molecular weight substances. Neurosci Lett (in press)

122. Weiss DG, Gross GW (1982) Intracelluar transport in axonal microtubular domains. I. Theoretical considerations on the essential properties of a force generating mechanism. Protoplasma (in press)

123. Weiss DG, Gross GW (1982) The microstream hypothesis: characteristics, predictions and compatibility with data. In: Weiss DG (ed) Axoplasmic transport. Springer, Berlin Heidelberg New York, pp 362–383

124. Weiss DG, Schmid G, Wagner L (1980) Influence of microtubule inhibitors on axoplasmic transport of free amino acids. Implications for the hypothetical transport mechanism. In: De Brabander M, De Mey J (eds) Microtubules and microtubule inhibitors 1980. Elsevier/ North Holland, Amsterdam, pp 31–41

125. Willard M, Cowan WM, Vagelos PR (1974) The polypeptide composition of intra-axonally transported proteins: evidence for four transport velocities. Proc Natl Acad Sci USA 71: 2183–2187

126. Zelená J, Lubińska L, Gutmann E (1968) Acccumulation of organelles at the ends of interrupted axons. Z Zellforsch Mikrosk Anat 91:200–219

127. Zimmermann H, Whittaker VP (1974) Effect of electrical stimulation on the yield and composition of synaptic vesicles from the cholinergic synapses of the electric organ of *Torpedo:* a combined biochemical, electrophysiological and morphological study. J Neurochem 22:435–450

Section 1
The Physiological Role of Axoplasmic Transport

Axoplasmic Transport and Synaptic Transmission

ANNICA DAHLSTRÖM [1]

The aim of this contribution primarily is to give a short review of studies which point to the close dependence on anterograde axonal transport of events, related to synaptic transmission.

Probably the first suggestion of the importance of passage of material from the cell body down the axons for the maintenance of stimulation-induced "synaptic transmission" was put forward by Scott in 1906 [15]. In fact he formulated thoughts on both axonal transport and chemical neurotransmission. In frogs he studied the effect of cutting the dorsal roots from the ganglia, on the fatiguability of the reflex arch following stimulation of the central end. After a large number of stimuli roots connected to their cell bodies could recover, while cut roots could not. This was not due to degeneration, because cut roots, if left unstimulated, could activate the reflex arch upon stimulation at a time when cut, stimulated roots were fatigued. Scott summarizes: " . . . it seems to me simpler to suppose that the nerve cells secrete a substance the passage of which from the nerve endings is necessary to stimulation [2]. The recovery of effect after transient fatigue I attribute to the passage of a portion of this substance down the nerve fibre to the nerve endings [3]." Thus, synaptic transmission is dependent upon anterograde axonal transport, a hypothesis expressed in an article published in 1906 [15].

Many studies on the dependence of transmission-related parameters have been performed in denervated tissues where results, following transection of the nerve close to the innervated tissue, were compared with those obtained if a length of nerve was left attached to the innervated organ. Thus, in *cholinergic* nerves, a fall in acetylcholine (ACh)-content, failure of miniature end-plate potentials (mepps) and of junctional transmission, occurrence of denervation alterations in skeletal muscle have been studied using this experimental model. It was found that if a longer stump of the cervical preganglionic nerve was attached, the onset of ACh decrease in the superior cervical ganglion (SCG) of rat was delayed by about 3 h if the nerve left in continuity with the ganglia was 10–14 mm [10]. Failure of mepps and junctional transmission in rat diaphragm occurred 45 min later for each additional cm of phrenic nerve left attached [13]. Appearance in rat extensor digitorum longus muscle fibres

---

1 Institute of Neurobiology, University of Göteborg, P.O. Box 33031, S-400 33 Göteborg, Sweden
2 i.e. chemical transmission (A.D.)
3 i.e. axonal transport (A.D.)

Axoplasmic Transport in Physiology and Pathology
(ed. by D.G. Weiss and A. Gorio)
© Springer-Verlag Berlin Heidelberg 1982

of tetrodotoxin resistant action potentials, characteristic for denervated muscles, was delayed for 2 h for each cm of nerve stump remaining [1]. In the *adrenergic* system the onset of noradrenaline (NA)-decrease in gastrochemic muscles following sectioning of the sciatic nerve (carrying postganglionic adrenergic nerves, innervating e.g. blood vessels in the muscle) was delayed 1.2 h for each cm of sciatic nerve left with the muscle [2]. Degeneration secretion in rat salivary glands (an indication of leakage of NA from disintegrating nerve terminals) started 2 h earlier if the axons were crushed near the hilus of the gland than if the crush was made close to the superior cervical ganglion (SCG), with approximately 15–20 mm nerve attached [1].

The above studies (and many others) thus seem to indicate that some substance(s), transported in the attached nerve stump to the nerve terminals at a rate of 5–10 mm/h, can maintain transmitter levels, mepps, junctional transmission, nerve terminal integrity, and prevent the effector cells from turning "denervated". On the other hand, the signal to the nerve terminals for transmission to cease and degenerative changes to start may be "positive" rather than "negative". Instead of assuming that some factor(s) necessary for maintenance is used up earlier with a short nerve stump than with a longer one, one may speculate on an "axotomy information" being carried down the axons, and arriving later to the nerve endings if the distance it has to travel is longer [12]. Some evidence for the existence of such an axotomy message comes from the observation that the degeneration contraction of the lower eye-lid of rat, induced by removal of the SCG and caused by leakage of NA from degenerating adrenergic nerve terminals, could be delayed for several hours if colchicine or vinblastine was given systemically to the animal 0–5 h after the axotomy, but not if the drugs were given later [12]. Assuming that the effect of systemic mitotic inhibitors is on the microtubules of the adrenergic nerves, resulting in impaired axonal transport in the attached nerve stump, this may point to the existence of some axotomy information substance. Local treatment of the SCG with millimolar concentrations of colchicine or vinblastine, however, which inhibits the fast anterograde transport of NA-granules and other constituents, caused degeneration contraction, starting 13–14 h after treatment, i.e. the same time as after axotomy [7, 12].

Instead of studying the disappearance of transmission-related parameters after interference with axonal transport, the recovery of transmission can be studied in an experimental condition where pharmacologic treatment has caused transmission failure. Such a situation is created for the adrenergic system following a large dose of reserpine (10 mg/kg i.p. to rats). This drug irreversibly impairs the NA-storage mechanism in the amine granules of the entire neuron, causing NA-levels to drop virtually nil, abolishing the capacity for uptake-retention of ^{3}H-NA given i.v., and results in a complete cessation of noradrenergic transmission 1–12 h after injection, as observed both behaviourally and electrophysiologically [4]. For recovery of adrenergic function of the animal an uninterrupted supply of newly synthesized amine granules is essential [6]. In Fig. 1 a summary of data concerning recovery of various functional parameters in long adrenergic nerves after reserpine is given. NA-fluorescence first appears in a perinuclear zone (site of production and package of amine granules of their precursors), then at a proximal and later at a distal level of the axons, and finally traces of NA can be observed in the nerve terminals (by biochemical and histochemical methods). As soon as a few percent of the endogenous NA has

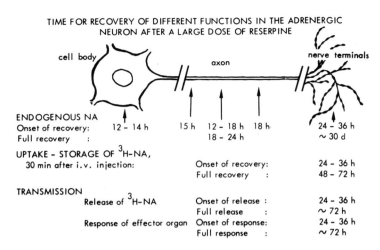

TIME FOR RECOVERY OF DIFFERENT FUNCTIONS IN THE ADRENERGIC
NEURON AFTER A LARGE DOSE OF RESERPINE

ENDOGENOUS NA
Onset of recovery: 12 – 14 h   15 h   12 – 18 h   18 h   24 – 36 h
Full recovery   :   18 – 24 h   ~ 30 d

UPTAKE – STORAGE OF ^{3}H-NA,
    30 min after i.v. injection: Onset of recovery: 24 – 36 h
                                             Full recovery   :   48 – 72 h

TRANSMISSION
        Release of ^{3}H-NA   Onset of release :   24 – 36 h
                                    Full release   :   ~ 72 h
    Response of effector organ   Onset of response: 24 – 36 h
                                    Full response   :   ~ 72 h

Fig. 1. Schematic illustration showing the times for onset of recovery and full recovery of various
parameters, important for transmission, in a long adrenergic neuron of the rat after one large dose
of reserpine (10 mg/kg i.p.). After reserpine these various parameters are virtually zero at 6–12 h.
(The figure is modified after Fig. 1 in [6], and references to the basic data are found in this
reference)

recovered in nerve terminals, the capacity for uptake-storage of ^{3}H-NA starts to
recover, and this is also the time when the first evidence for a returning response of
the effector organ to sympathetic stimulation occurs. Evidently, new functioning
amine-storing granules have reached the terminals and can function for storage, uptake
and release of NA. Much evidence supports the view that the "young" amine granules
are particularly important for transmission; recovery of uptake-retention of ^{3}H-NA
and transmission is complete when endogenous NA-levels are only recovered by about
10%. The axonal granules are rich in dopamine-β-hydroxylase (DBH), and the results
of Brimijoin [3], who calculated the turn-over of DBH in adrenergic nerve terminals
of the rat hindleg, fits very well with our estimated half-life of "young amine granules"
in the same system [6]. The other type of NA-granules in the nerve terminals are
smaller, lighter and contain little DBH in relation to NA, and may form from the
larger "young" granules [8, 16].

    If the axonal transport of newly formed NA granules is blocked during recovery
after reserpine, the onset of recovery of the above mentioned parameters is delayed.
In experiments where the contraction of the rats lower eye-lid following preganglionic
stimulation of the SCG was followed after reserpine, local treatment of one ganglion
with 1 mM vinblastine markedly delayed the return of adrenergic transmission in the
ipsilateral eye-lid smooth muscle (Fig. 2) [9]. In the adrenergic system, therefore,
a close correlation between axonal transport and adrenergic transmission seems established.

    In the pigeon optic system, Perísic and Cuénod [14] could depress the recorded
postsynaptic potential in optic tectum after stimulation of the optic nerve, by local
injection into the contralateral eye of colchicine. Colchicine did not interfere with

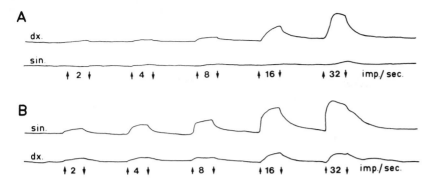

Fig. 2. Contraction recording of the rat lower eye-lid in response to preganglionic stimulation of the superior cervical ganglion (SCG) at various frequencies. **A** from a rat given reserpine (10 mg/kg i.p.), 23 h before local treatment of the SCG with vinblastine (1 mM) on the left, and saline on the right side. The response to stimulation was recorded 10 h later, i.e. a total of 33 h after the reserpine injection. Note that with 8−32 Hz a clear contractile response was recorded on the control (saline) side, while no clear response was observed on the vinblastine treated side. **B** from another rat, treated with vinblastine (1 mM) locally on the right SCG and with saline (control) on the left side 9 h before the stimulation experiment. Reserpine was given 24 h prior to SCG treatments, i.e. 33 h before the stimulation experiment. The control side gave a clear response to stimulation at 4−32 Hz, but the vinblastine treated side showed a weak contraction at 16−32 Hz. In **B** the SCG was vinblastine treated 24 h after reserpine but in **A** 23 h after reserpine, a fact which may contribute to the differences in response to stimulation between rat **A** and rat **B**. Thus, during the 24th hour after reserpine more newly formed functioning amine granules may have been passed down the axon before the vinblastine block of transport [9]

nerve impulses of optic nerve or with protein synthesis in the retinal ganglion cells. Axonal transport, however, was blocked, but the effect on both axonal transport, nerve terminal morphology and postsynaptical tectal potentials [5] were reversible, and recovery of transmission recurred by 2 weeks with 10 μg and by 8 weeks after 100 μg of colchicine. From this very elegant experiment the authors conclude: " . . . our results would be consistent with the notion that material, which is normally provided by the ganglion cell body to nerve terminals, and which may be arrested by this drug (colchicine), plays an essential role in the maintenance of synaptic functions" [14].

Acknowledgments. The work from the author's laboratory described in this section was supported by grants from the Swedish MRC (14X-2207; 04P-4173), by M. Bergwall's Foundation and W. and M. Lundgren's Foundation.

References

1. Almgren O, Dahlström A, Häggendal J (1976) Degeneration secretion and noradrenaline disappearance in rat salivary glands following proximal or distal axotomy. Acta Physiol Scand 98:457–464
2. Bareggi S, Dahlström A, Häggendal J (1974) Intra-axonal transport and degeneration of adrenergic nerve terminals after axotomy with a long and short nerve stump. Med Biol 52: 327–335
3. Brimijoin S (1972) Transport and turnover of dopamine-β-hydroxylase (EC 1.14.12.1) in sympathetic nerves of the rat. J Neurochem 19:2183–2193
4. Carlsson A (1965) Drugs which block the storage of 5-hydroxytryptamine and related amines. In: Erspamer V (ed) Handbuch der experimentellen Pharmakologie, vol XIX. Springer, Berlin Heidelberg New York, pp 529–592
5. Cuénod M, Boesch J, Marko P, Perísic M, Sandri C, Schonbach J (1972) Contributions of axoplasmic transport to synaptic structures and functions. Int J Neurosci 4:77–87
6. Dahlström A, Häggendal J (1973) Intra-axonal transport of young amine granules: Implications for nerve terminal function. In: Genazzani E, Herken H (eds) Central nervous system studies on metabolic regulation and function. Springer, Berlin Heidelberg New York, pp 94–103
7. Häggendal J, Dahlström A (1971) The functional role of amine storage granules in the sympatho-adrenal system. In: Subcellular organization and function in endocrine tissues. Mem Soc Endocrinol 19:651–667
8. Häggendal J, Dahlström A (1971) The importance of axoplasmic transport of amine granules for the functions of adrenergic neurons. Acta Neuropathol Suppl V:238–248
9. Häggendal J, Dahlström A (unpublished data)
10. Häggendal J, Dahlström A, Saunders N (1973) Axonal transport and acetylcholine in rat preganglionic neurons. Brain Res 58:494–499
11. Harris JB, Thesleff S (1972) Nerve stump length and membrane changes in denervated skeletal muscle. Nature (London) New Biol 236:60–61
12. Lundberg D (1972) Effects of colchicine, vinblastine and vincristine on degeneration transmitter release after sympathetic denervation studied in the conscious rat. Acta Physiol Scand 85:91–98
13. Miledi R, Slater CR (1970) On the degeneration of rat neuromuscular junctions after nerve section. J Physiol (London) 207:507–528
14. Perísic M, Cuénod M (1972) Synaptic transmission depressed by colchicine blockade of axoplasmic flow. Science 175:1140–1142
15. Scott FH (1906) On the relation of nerve cells to fatigue of their nerve fibres. J Physiol (London) 34:145–162
16. Smith AD, Winkler H (1972) Fundamental mechanisms in the release of catecholamines. In: Blaschko H, Winkler H (eds) Handbuch der experimentellen Pharmakologie, vol 33: Catecholamines. Springer, Berlin Heidelberg New York, pp 573–617

Effect of Physiological Activity in Goldfish Optic Axons on Axonal Transport of Protein and Nucleosides

BERNICE GRAFSTEIN and D. LOUISE EDWARDS [1]

Introduction

Among its various contributions to neuronal function, axonal transport plays an essential role in the maintenance of physiological activity, primarily by supplying constituents required for synaptic transmission [6, 7]. It might therefore be anticipated that physiological activity would in turn have an influence on axonal transport. There is ample evidence that the *velocity* of fast axonal transport is unaffected by neuronal activity [7], but is there an effect on the *amount* of material that is transported? Such an effect might operate, for example, by a change in the synthesis or degradation of transportable material, or by a change in the proportion of available material that is loaded into the transport system, the loading being a process that has been shown to be regulated by a Ca^{++}-dependent step [8, 9]. Thus far there have been only a few reports of alterations in the amount of fast-transported proteins in response to alterations in physiological activity [2, 16, 19, 22] or to electrical stimulation [14]. Some observations on the axonal transport of adenosine, on the other hand, suggest that physiological activity may have a significant influence ([21]; see also Schubert and Kreutzberg, this volume).

Studies on Goldfish Optic Axons

In our own approach to the problem of the role of physiological activity on axonal transport, we have taken advantage of the properties of the goldfish visual system that have made it a convenient object for axonal transport experiments [4]. Axonal transport of various materials in the optic axons can be studied by injecting a radioactively labeled precursor into the eye and measuring the amount of labeled material subsequently appearing in the optic nerve and tectum. In order to vary the physiological activity in the optic axons, we have suppressed visual activity by intraocular injection of tetrodotoxin (TTX), and determined the differences between axonal transport

---

1 Department of Physiology, Cornell University Medical College, New York, New York 10021, USA

Axoplasmic Transport in Physiology and Pathology
(ed. by D.G. Weiss and A. Gorio)
© Springer-Verlag Berlin Heidelberg 1982

under these conditions and under conditions of normal visual activity. Some of these experiments have been briefly described in [6].

Axonal Transport of Protein

No differences in protein transport were seen up to 8 days after the intraocular injection of radioactively labeled proline (Fig. 1A, B), a time encompassing the arrival of fast transport in the optic nerve and tectum, and the penetration of slow transport into the optic nerve [15]. Maintained inactivity produced by repeated intraocular injection of TTX beginning 6 days before the injection of labeled proline likewise had

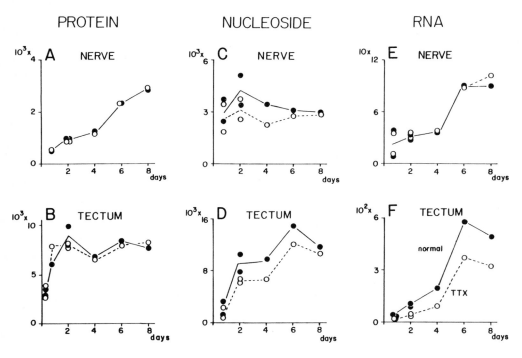

Fig. 1. Axonally transported labeled protein (**A, B**), nucleoside (**C, D**) and RNA (**E, F**) in optic nerve *(upper panels)* and optic tectum *(lower panels)* following injection of ^{14}C-proline (**A, B**) or ^{3}H-adenosine (**C, D, E, F**) into one eye of normal fish (●———●) and fish in which visual activity had been eliminated by intraocular injection of TTX (o - - - - - o). Animals in each normal group were injected in the right eye with 4 μl of citrate buffer and animals in each TTX group with 0.07 μg TTX in citrate buffer at least 1 h before injection of label into the same eye. At appropriate times thereafter the fish were decapitated and the optic nerves and tecta dissected out. Each tissue sample was placed in 0.5 ml cold 10% trichoroacetic acid (TCA) for 24 h, then transferred to 0.3 ml of tissue solubilizer and allowed to dissolve. The TCA extract and the dissolved tissue were each prepared for liquid scintillation counting. In each animal, the difference in radioactivity between the left and right sides represents the amount of labeled material arriving along the axons from the right eye [15]. **A, B, E**, and **F** represent the TCA-insoluble fraction of the labeled tissue, **C, D** the TCA-soluble fraction. Ordinate values are in dpm per μCi injected per mm length of nerve or per mg dry weight of tectum. Each *point* is the mean value from 8–10 fish

no influence on axonal transport of protein for up to 4 days after the proline injection. There was also no effect of activity demonstrable when transported proteins were labeled with other amino acids, including lysine, methionine, or asparagine (which would presumably be less extensively reutilized than proline [10]) or with fucose (which would be incorporated specifically into glycoproteins [3]). We have other evidence showing that physiological activity has no effect on velocity or amount of slow transport [5, 6].

The absence of any effect of physiological activity on fast protein transport is surprising in view of the fact that, at least in noradrenergic neurons, the transport of enzymes involved in transmitter metabolism appears to change in amount with activity [19, 22]. However, it is possible that in the goldfish optic axons transmitter-related materials make only a small contribution to fast transport.

Axonal Transport of Nucleosides

Studies corresponding to those described above for protein were carried out with intraocular injection of labeled adenosine. In this case the transported labeled trichloroacetic acid (TCA)-soluble fraction, which contains the nucleoside and its related metabolic derivatives [11], showed no change in transport velocity, but the amount of this material appearing in the optic nerve and tectum was up to 50% higher when visual activity was normal than when it was inactivated with TTX (Fig. 1C, D).

In addition to TCA-soluble material, there is a considerable amount of labeled RNA (TCA-insoluble) appearing along the optic axons [11]. Under the conditions of the present experiment, this RNA is unlikely to be transported [12], since it is evident very soon after the precursor injection. It presumably originates from labeled transported nucleoside (or nucleotide) which has been released from the axons and become incorporated into RNA in adjacent cells, i.e., glial cells of the optic nerve and glial cells and postsynaptic neurons of the optic tectum [1, 13, 17]. We found that the amount of labeled RNA appearing in the optic nerve was identical in both the normal (visually active) and TTX-injected (visually inactivated) animals (Fig. 1E). We must assume, therefore, that the amount of labeled RNA precursors released from the optic axons is not affected by activity. Thus the released material apparently originates from only a limited (activity-insensitive) fraction of the labeled TCA-soluble material in the axon.

In contrast to the observations in the optic nerve, we found that there was a large activity-related difference in the amount of labeled RNA appearing in the optic *tectum,* presumably having been synthesized in the postsynaptic neurons from precursors released from the presynaptic axons. The amount in the normal animals was up to 100% more than in the TTX-injected animals (Fig. 1F). Evidence that the activity-related difference in labeled RNA in the tectum was due to a difference in incorporation into postsynaptic tectal neurons was obtained by determining the distribution of the labeled RNA in autoradiograms of the tectum. The largest differences were found in the tectal layers containing cell bodies of neurons receiving input from the optic axons (Fig. 2).

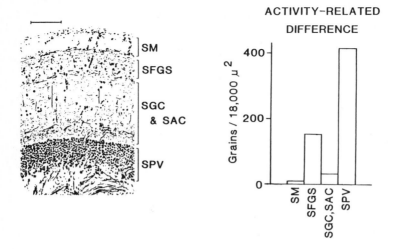

Fig. 2. Analysis of autoradiograms of optic tecta of fish injected with ³H-adenosine in one eye. *Left,* silver-stained section of goldfish optic tectum to show tectal layers from which grain counts were made. *SM* stratum fibrosum marginale; *SFGS* stratum fibrosum et griseum superficiale; *SGC* and *SAC* stratum griseum centrale and stratum album centrale; *SPV* stratum periventriculare. *Bar* 100 μm. *Right,* activity-related differences in grain densities of various tectal layers. The values shown represent the differences between grain counts in tecta of normal fish (injected with citrate buffer in one eye) and fish injected with TTX in one eye before injection of ³H-adenosine into the same eye (see legend to Fig. 1 for details of injections). In preparation of the tissues for autoradiography [15], fixation with Bouin's solution was used, so that the radioactivity recorded represents labeled RNA

Activity-related differences in the tectum similar to those obtained with ³H-adenosine were obtained with ¹⁴C-adenosine, with ¹⁴C-uridine, and with ³H-guanosine. It is possible that the increase in transported radioactivity may represent an increase in specific activity of the labeled material, e.g. due to a change in the relative proportions of exogenous and endogenous precursor pools contributing to the transport stream [23]. However, the relatively uniform changes that we found with various nucleosides irrespective of differences that presumably exist in the sizes of their intracellular pools and in their rates of turnover, suggests that the observed changes represent net changes in the amounts transported. In each case the activity-related difference for RNA was greater than the activity-related difference for TCA-soluble material, which implies that there is an activity-related increase in the amount of nucleoside incorporated into RNA by the postsynaptic neurons independent of whether or not there is any increase in the amount of nucleoside transported in the presynaptic axons or released from them

Activity-related changes in axonal transport and transcellular transfer of nucleoside may have significance for trophic interactions among neurons or between neurons and Schwann or glial cells. One possibility is that the transported nucleoside may be a preferential source of precursors for RNA synthesis in the cells to which it is made available [18]. Another possibility is that the transferred nucleoside may play a role in regulating the synthesis of cyclic nucleotide which normally serves as a "second messenger" for synaptic activation ([20]; see also Schubert and Kreutzberg, this volume). The similarity of the activity-related changes that we observed with various nucleosides seems to provide some support for the first of these hypotheses.

Summary

The normal level of physiological activity in goldfish optic axons has no effect on the overall amount of proteins in fast axonal transport or the transport velocity. However the amount of various transported nucleosides is about 50% higher in normally active axons than in the absence of activity. Also, the incorporation of transported nucleosides into RNA in postsynaptic neurons in the optic tectum is about 100% higher in the normally active system, whereas there is no activity-related difference in the incorporation into RNA in the glial cells of the optic nerve.

Acknowledgments. We are grateful to Stephen Lewis and Roberta M. Alpert for technical assistance, and to Maureen McEntee for typing the manuscript. This research was supported by USPHS grants NS-09015 from NINCDS and EY-02696 from NEI.

References

1. Autilio-Gambetti L, Gambetti P, Shafer B (1973) RNA and axonal flow. Biochemical and autoradiographic study in the rabbit optic system. Brain Res 53:387–398
2. Dahlström A, Heiwall P-O, Bööj S, Dahllöf A-G (1978) The influence of supraspinal impulse activity on the intra-axonal transport of acetylcholine, choline acetyltransferase and acetylcholinesterase in rat motor neurons. Acta Physiol Scand 103:308–319
3. Forman DS, Grafstein B, McEwen BS (1972) Rapid axonal transport of [^{3}H]fucosyl glycoproteins in the goldfish optic system. Brain Res 48:327–342
4. Grafstein B (1975) The eyes have it: axonal transport and regeneration in the optic nerve. In: Tower DB (ed) The nervous system, vol I: The basic neurosciences. Raven Press, New York, pp 147–151
5. Grafstein B, Alpert RM (1982) Properties of slow axonal transport: studies in goldfish optic axons. In: Weiss DG (ed) Axoplasmic transport. Springer, Berlin Heidelberg New York
6. Grafstein B, Edwards DL, Alpert RM (1981) Axonal transport and neuronal activity. In: Schweiger HG (ed) International cell biology 1980–1981. Springer, Berlin Heidelberg New York, pp 728–736
7. Grafstein B, Forman DS (1980) Intracellular transport in neurons. Physiol Rev 60:1167–1283
8. Hammerschlag R (1980) The role of calcium in the initiation of fast axonal transport. Fed Proc 39:2809–2814
9. Hammerschlag R (1982) Multiple roles of calcium in the initiation of fast axonal transport. In: Weiss DG (ed) Axoplasmic transport. Springer, Berlin Heidelberg New York
10. Heacock AM, Agranoff BW (1977) Reutilization of precursor following axonal transport of [^{3}H]proline-labeled protein. Brain Res 122:243–254

11. Ingoglia NA, Grafstein B, McEwen BS, McQuarrie IG (1973) Axonal transport of radioactivity in the goldfish optic system following intraocular injection of labeled RNA precursors. J Neurochem 20:1605–1615

12. Ingoglia NA, Weis P, Mycek J (1975) Axonal transport of RNA during regeneration of the optic nerves of goldfish. J Neurobiol 6:549–564

13. Ingoglia NA, Zanakis MF (1982) Axonal transport of 4S RNA. In: Weiss DH (ed) Axoplasmic transport. Springer, Berlin Heidelberg New York, pp 161–169

14. Lux HD, Schubert P, Kreutzberg GW, Globus A (1970) Excitation and axonal flow: autoradiographic study on motoneurons intracellularly injected with a ^{3}H-amino acid. Exp Brain Res 10:197–204

15. McEwen BS, Grafstein B (1968) Fast and slow components in axonal transport of protein. J Cell Biol 38:494–508

16. Norström A, Sjöstrand J (1972) Effect of salt-loading, thirst and water-loading on transport and turnover of neurohypophysial proteins of the rat. J Endocrinol 52:87–105

17. Peterson JA, Bray JJ, Austin L (1968) An autoradiographic study of the flow of protein and RNA along peripheral nerve. J Neurochem 15:741–745

18. Politis MJ, Ingoglia NA (1979) Axonal transport of nucleosides, nucleotides and 4S RNA in the neonatal rat visual system. Brain Res 169:343–356

19. Reis DJ, Ross RA, Gilad G, Joh TH (1978) Reaction of central catecholaminergic neurons to injury: model systems for studying the neurobiology of central regeneration and sprouting. In: Cotman C (ed) Neuronal plasticity. Raven Press, New York, pp 197–226

20. Schubert P, Kreutzberg GW (1974) Axonal transport of adenosine and uridine derivatives and transfer to postsynaptic neurons. Brain Res 76:526–530

21. Schubert P, Lee K, West M, Deadwyler S, Lynch G (1976) Stimulation-dependent release of ^{3}H-adenosine derivatives from central axon terminals to target neurons. Nature (London) 260:541–542

22. Thoenen H, Mueller RA, Axelrod J (1970) Phase difference in the induction of tyrosine hydroxylase in cell body and nerve terminals of sympathetic neurones. Proc Natl Acad Sci USA 65:58–62

23. Whitnall MH, Grafstein B (1981) The relationship between extracellular amino acids and protein synthesis is altered during axonal regeneration. Brain Res 220:362–366

Fate of Axonally Transported Proteins in the Nerve Terminal

MATS SANDBERG, ANDERS HAMBERGER, INGEMAR JACOBSSON,
and JAN-OLOF KARLSSON [1]

Knowledge about the constituents of the axonal transport has increased considerably during the last years. For example, the major structural elements in the axon are transported at a slow rate [8, 11, 25]. The half-life for proteins of this phase has been found to be in the order of 1 to 4 weeks [3, 10]. The fast phase carries a heterogeneous group of proteins [12, 23] which have been shown to turnover more rapidly than the slowly transported proteins. At least some components appear to have half-lives in the order of a few hours [5, 10, 24]. Except for these rough estimations of turnover rates, little is known about the fate of axonally transported proteins in the terminal. With the exception of those components that are carried backwards via the retrograde axonal transport, the remaining proteins in the terminal have to be degraded and/or secreted to counterbalance the axonal transport. The quantitative significance of axonal transport leads to the conclusion that a considerable portion of neuronal macromolecules are degraded outside the nerve cell body. The fraction of small molecular weight compounds which may be formed during degradation of axonally transported proteins cannot be used locally as protein precursors, since a protein synthesis machinery is largely missing in the axon terminal. In cholinergic or adrenergic neurons it is commonly thought that the content of the transmitter-containing vesicles is released via exocytosis during nerve activity. Similarly, neurosecretory neurons release their axonally transported hormones and other substances upon functional stimulation [2, 16].

Other non-cholinergic and non-adrenergic neurons, for example peptidergic and neurons using amino acid transmitters, might also secrete material derived from axonally transported material. Rapidly transported proteins have been found to be released from frog sciatic nerve in vitro [7]. However, in a recent report using a similar systems contradictory results were reported [22]. Substances, which have an innervation-like effect on organ-cultured rat muscles, have been shown to be axonally transported and released from nerve during stimulation [26].

It has recently been shown that rapidly and slowly transported proteins from retinal ganglion cells of the rabbit in an in vitro system may be released from the terminal region upon chemical depolarization (Fig. 1) [18, 19]. However, it seems likely that a major part of the rapidly transported proteins must be degraded in the nerve terminals. It is possible that Ca-ions in the nerve terminal is of importance in this

1 Institute of Neurobiology, University of Göteborg, Box 33031, S-400 33 Göteborg, Sweden

Axoplasmic Transport in Physiology and Pathology
(ed. by D.G. Weiss and A. Gorio)
© Springer-Verlag Berlin Heidelberg 1982

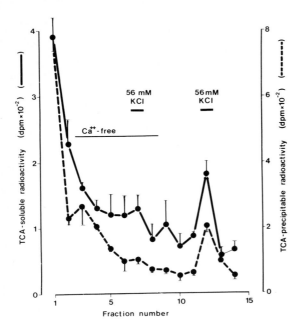

Fig. 1. Release of acid-soluble (——) and acid-insoluble (- - - -) radioactivity from slices of superior colliculus labelled by an intraocular injection of ^{3}H-glycine 18 h earlier. Pulsing with 56 mM KCl-66 mM NaCl was performed either in Ca-free media (supplemented with 10 mM $MgCl_2$) *(fraction 7)*, or in normal 1.2 mM Ca *(fraction 12)* as indicated

degradation process. Ca^{2+} may promote disassembly of microtubules and activate proteases which specifically act upon microtubule-associated proteins [20]. A Ca^{2+}-activated protease in the axoplasm which selectively degrades neurofilament proteins has recently been described [17]. Lasek and Hoffman [14] have made the very interesting suggestion that an activation of presynaptic proteases and a selective degradation of cytoskeletal elements are dependent on the interaction of the axon terminal with the postsynaptic cell. Axonal growth and regeneration would thus be a consequence of a lack of contact with postsynaptic cells.

In a recent study of the superior colliculus it was shown [18, 19] that slowly and fast transported proteins from the retinal ganglion cells were degraded to acid-soluble components in the nerve terminal. A perfused tissue slice of the superior colliculus responded to depolarization with a Ca^{2+}-dependent release of acid soluble material (Fig. 1) [18, 19].

The release of acid-soluble radioactivity from slowly transported proteins has also been used to determine the rate of degradation of such proteins [15]. The release of transported glutamate [13] and of adenosine-derivatives [21] has been shown.

Ca^{2+}-dependent release of transported acid-soluble protein derivatives upon depolarization of a terminal region have, to our knowledge, not been demonstrated earlier. It is possible that a part of the released material may have physiological functions serving as transmitting or neuromodulating substances. The regulation of proteolytic activity generating such molecules in the nerve terminal may then be of crucial importance. The slice experiments also suggested a continuous replenishment of acid-soluble material into a releasable pool. These results are in agreement with suggestions concerning axonal transport made by Jones and McIlwain [9]. The degradation of rapidly

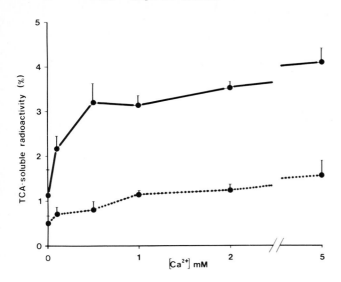

Fig. 2. Effects of Ca-concentration on the production of acid-soluble radioactivity from rapidly (——) and slowly (- - - - -) transported proteins, labelled by intraocular ^{3}H-glycine injections. The experiments were performed using a combined homogenate of the superior colliculus and the lateral geniculate body incubated at 37°C for 1 h in different Ca-concentrations

transported proteins to low molecular weight components was found to be highly Ca^{2+}-dependent (Fig. 2) and inhibited by sulfhydryl blockers when assayed in a homogenate system. Gel filtration of the soluble components derived from rapidly transported proteins in the presence of calcium has shown the formation of peptides and amino acids [19]. These findings imply that axonally transported proteins (rapidly and slowly transported) are a source of amino acids and/or peptides in the nerve ending and these substances could consequently be of potential importance in neurotransmission.

The breakdown of acid-soluble material was stimulated by Ca^{2+} at neutral pH. The quite high Ca^{2+}-concentration (0.1 mM−0.5 mM) needed to activate such proteases [6, 17, 19] is probably not generally reached in the terminal even during depolarization. However, as pointed out by Baudry and Lynch [1], it is not unlikely that higher Ca^{2+}-concentrations could be reached during repetitive stimulation.

It is thus possible that nerve activity with an increase in intraterminal Ca^{2+}-concentration [4] could stimulate some proteolytic activities in the nerve terminal. The newly formed low molecular weight components may then be released from the terminal and subserve important physiological functions.

Acknowledgments. We thank the Swedish Medical Research Council (grants no. 00164 and 05932) and Torsten and Ragnar Söderbergs Stiftelse, and Axel and Margaret Ax:son-Johnsons Stiftelse.

References

1. Baudry M, Lynch G (1980) Regulation of hippocampal glutamate receptors: Evidence for the involvement of a calcium-activated protease. Proc Natl Acad Sci USA 77:2298–2302
2. Berry RW (1979) Secretion of axonally transported neural peptides from the nervous system of *Aplysia*. J Neurobiol 10:499–508
3. Black MM, Lasek RJ (1979) Axonal transport of actin: slow component b is the principal source of actin for the axon. Brain Res 171:401–413
4. Blaustein MP, Ratzlaff RW, Kendrick NK (1978) The regulation of intracellular calcium in presynaptic nerve terminals. Ann NY Acad Sci 307:195–211
5. Goodrum JF, Toews AD, Morell P (1979) Axonal transport and metabolism of [^{3}H]-fucose- and [^{35}S]-sulfate-labelled macromolecules in the rat visual system. Brain Res 176: 255–272
6. Guroff G (1964) A neutral, calcium-activated proteinase from the soluble fraction of the rat brain. J Biol Chem 239:149–155
7. Hines JF, Garwood MM (1977) Release of protein from axons during rapid axonal transport: an in vitro preparation. Brain Res 125:141–148
8. Hoffman PN, Lasek RJ (1975) The slow component of axonal transport. Identification of major structural polypeptides of the axon and their generality among mammalian neurons. J Cell Biol 66:351–366
9. Jones DA, McIlwain H (1971) Amino acid production and translocation in incubated and superfused tissues from the brain. J Neurobiol 2:311–326
10. Karlsson J-O, Sjöstrand J (1971) Synthesis, migration and turnover of protein in retinal ganglion cells. J Neurochem 18:749–767
11. Karlsson J-O, Sjöstrand J (1971) Transport of microtubular protein in axons of retinal ganglion cells. J Neurochem 18:875–984
12. Karlsson J-O, Sjöstrand J (1971) Characterization of the fast and slow components of axonal transport in retinal ganglion cells. J Neurobiol 2:135–143
13. Kerkut GA, Shapira A, Walker RJ (1967) The transport of ^{14}C-labelled material from CNS ⇌ muscle along a nerve trunk. Comp Biochem Physiol 23:729–748
14. Lasek RJ, Hoffman PN (1976) The neuronal cytoskeleton, axonal transport and axonal growth. In: Goldman R, Pollard T, Rosenbaum J (eds) Cell motility. Cold Spring Harbor Lab, New York, pp 1021–1049
15. Nixon R (1980) Protein degradation in the mouse visual system. I. Degradation of axonally transported and retinal proteins. Brain Res 200:69–93
16. Norström A, Sjöstrand J (1971) Effect of haemorrhage on the rapid axonal transport of neurohypophysial proteins in the rat. J Neurochem 18:2017–2026
17. Pant HC, Gainer H (1980) Properties of a calcium-activated protease in squid axoplasm which selectively degrades neurofilament proteins. J Neurobiol 11:1–12
18. Sandberg M, Hamberger A, Karlsson J-O, Tirillini B (1980) Potassium stimulated release of axonally transported radioactivity from slices of rabbit superior colliculus. Brain Res 188: 175–183
19. Sandberg M, Hamberger A, Jacobsson I, Karlsson J-O (1980) Role of calcium ions in the formation and release of low-molecular weight substances from optic nerve terminals. Neurochem Res 5:1179–1192
20. Sandoval IV, Weber K (1978) Calcium-induced inactivation of microtubule formation in brain extracts. Eur J Biochem 92:463–470
21. Schubert P, Lee K, West M, Deadwyler S, Lynch G (1976) Stimulation-dependent release of ^{3}H-adenosine derivates from central axon terminals to target neurons. Nature (London) 260:541–542
22. Tedeschi B, Wilson DL, Zimmerman A, Perry GW (1981) Are axonally transported proteins released from sciatic nerves? Brain Res 211:175–178

23. Wagner JA, Kelly AS, Kelly RB (1979) Nerve terminal proteins of the rabbit visual relay nuclei identified by axonal transport and two-dimensional gel electrophoresis. Brain Res 168:87–117
24. Willard M, Cowan WM, Vagelos PR (1974) The polypeptide composition of intra-axonally transported proteins: Evidence for four transport velocities. Proc Natl Acad Sci USA 71: 2183–2187
25. Willard M, Wiseman M, Levine J, Skene P (1979) Axonal transport of actin in rabbit retinal ganglion cells. J Cell Biol 81:581–591
26. Younkin SG, Brett RS, Davey B, Younkin CH (1978) Substances moved by axonal transport and released by nerve stimulation have an innervation-like effect on muscle. Science 200: 1292–1295

Transneuronal Transport: A Way for the Neuron to Communicate with Its Environment

PETER SCHUBERT and GEORG W. KREUTZBERG[1]

Molecular transfer between the neuron and elements of its environment has become a favoured research topic in recent years although its existence has only been established for a short time. A trans-synaptic transfer of radioactive proteins at neuromuscular junctions was first suggested, but not clearly demonstrated, by the low resolution autoradiographs of Korr et al. in 1967 [14]. In 1971 Bernice Grafstein demonstrated transfer of radioactivity in the mammalian visual system [10]. After injection of ^{3}H-amino acids into the eye she found radioactivity not only in the lateral geniculate nucleus but most surprisingly, also in the visual cortex.

This discovery was confirmed subsequently in a number of different systems, e.g. in peripheral nerve fibers where transfer to the target cells was observed [1, 7, 22]. Further detailed studies by Grafstein et al. revealed that transneuronal transfer was not necessarily a trans-synaptic transfer [11, 40, 41]. Radioactive material was also released along the fiber tracts and was seen in regions contiguous but not synaptically connected to the target cell area of the brain. Release of material occurs of both the fast and slow axonal transport components. The nature of the material transferred is not yet clear. However, it is assumed that radioactive proteins undergo degradation as they are transported down the axon. The presence of proteolytic enzymes in synaptosomal fractions would favour the view that proteinaceous macromolecules are further degraded in the axonal terminals, leading to a release of low molecular weight material, e.g. amino acids, into the extracellular space. Uptake of these substances as part of the salvage economy of the brain tissue seems to occur, leading to a rather unspecific labeling of cells and structures of this region.

Molecular transfer occurs not only from the axon but also from the dendritic tree in which, similar to the axon, an effective transport system operates (Fig. 1). Such a transport of neuroplasm in the dendrites was first postulated on the basis of the cellulifugal distribution of radioactive proteins with time following a systemic application of labeled precursors [6]. A direct demonstration was possible by means of single cell injection with radioactive substances followed by autoradiography [9, 20, 36]. With glycine as precursor, protein transport has been seen to take place, without a lag phase, at a rate of approximately 1 μm/s. That means that within 15–20 min newly synthesized proteins can reach the whole dendritic arborization which in large

---

1 Max Planck Institute for Psychiatry, Kraepelinstraße 2, D-8000 München 40, Fed. Rep. of Germany

Axoplasmic Transport in Physiology and Pathology
(ed. by D.G. Weiss and A. Gorio)
© Springer-Verlag Berlin Heidelberg 1982

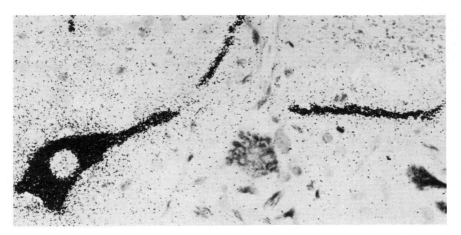

Fig. 1. Dendritic transport and release from dendrites as seen after intrasomal iontophoresis of
[3]H-choline into a cat spinal motoneuron. Magnified × 500. Autoradiography, counterstained
with toluidine-blue

motoneuron extends to approximately 1000 μm. The estimate of this rate is conservative since the experimental conditions do not allow extremely short term experiments to be performed and time has to be allowed for the iontophoretic tracer application [30].

Colchicine, which is known to block axonal transport by interfering with microtubular proteins, was also found to block dendritic transport [34]. Ultrastructural studies of the colchicine treated neurons with impaired dendritic transport revealed an enormous accumulation of filamentous tangles in the perikarya and the stem dendrites. Also a peculiar staining was seen in the Nissl preparations which may be due to a disorganization of the granular endoplasmic reticulum visible in the electron micrographs. After intracellular injection of isotopes the colchicine treated motoneurons displayed no protein transport to the peripheral dendrites, only some of the stem dendrites received newly synthesized proteins. Thus, it was concluded that destruction of the microtubular system also impaired dendritic transport.

In recent years it became clear that the dendrites also released substances in a hitherto unknown secretory process. In the single cell injection experiments it has been observed that radioactive material can also leave the motor neurons [30]. In contrast to experiments with ^{3}H-glycine, where radioactivity is seen exclusively within the injected neuron, a release from the dendrites occurs following injection of ^{3}H-choline, ^{3}H-fucose or ^{35}S-sulfate (Fig. 1). It therefore seems likely that some phospholipids, glycoproteins and acid mucopolysaccharides pass the dendritic membrane and become incorporated into structures in the environment of the injected neuron.

This phenomenon, originally described as "dendritic secretion" attracted considerable attention recently when dopamine was found to be liberated from the dendrites of striatal neurons [23]. Even compounds of high molecular weight are known to be released from these structures, e.g. the enzyme acetylcholinesterase. Thus, under certain experimental conditions, e.g. in the recovery phase after DFP intoxication or following axotomy of motor neurons, this enzyme has been shown electromicroscopically to increase in the extracellular space and to become very prominent in the basal lamina of local capillaries. Since the only visible source for AChE production in the CNS is the neuron, it has been assumed that it is the neuronal enzyme which accumulates at the capillaries [17]. This also demonstrates a pathway for molecular communication between the neurons and the local capillaries (see also [33]).

A further example for a release of macromolecules from neurons or dendrites has been demonstrated in the transfer of tetanus toxin from the postsynaptic to the presynaptic side [39]. After retrograde transport from the periphery towards the nerve cell soma, tetanus toxin was found also in boutons of afferent fibers inserting at the soma or the dendrites. Thus, retrograde transport and release from dendrites may be regarded as a route also for pathogenic compounds to be channelled within the CNS and to reach their targets.

About the functional significance of the observed transfer we can only speculate. It might well have to do more with signaling in terms of trophic effects than with simple supply of precursor substances for protein synthesis. In any case, these findings classify the nerve cell among the family of secretory cells; it appears conceivable that the neuron which underwent an extreme specialization during development still makes use of this hereditary capacity in order to complement its array of signals for intercellular communication. Accordingly, the concept has evolved that the neuron is not only using one principal transmitter but a mosaic of molecular signals for differential interaction with its environment. The routes along which these compounds are distributed upon release from neurons are determined by free diffusion in the extracellular space and by cellular re-uptake. Consequently, these compounds may reach a variety of target structures exchanging a variety of types of information.

In this chapter we will try to elucidate this concept by describing the available experimental data concerning the intracerebral channelling and function of adenine nucleotides. These compounds are transported and liberated from the neuron along the routes described; they are apparently able to exert a number of effects on manifold targets and may therefore function as additional neuronal signals by which the neuron communicates with its environment.

Axonal Transport and Release from Axons

Axonal transport of adenine compounds is well established. There is a fast transport of tetrachloroacetic acid (TCA)-soluble material down the axon to reach the axon terminals [2]. Here, some of the transported material is released from the axon terminals into the extracellular space and is usually found to be taken up by neighbouring structures such as the postsynaptic neurons and glial cells (Fig. 2, 3). Such a release of adenine compounds from axon terminals has been demonstrated in several pathways of the central nervous system, including the cortico-thalamic projection [29, 31], pathways originating in the somatosensory cortex, thalamocortical pathways [42] and the retinotectal and tecto-thalamic fiber system [12].

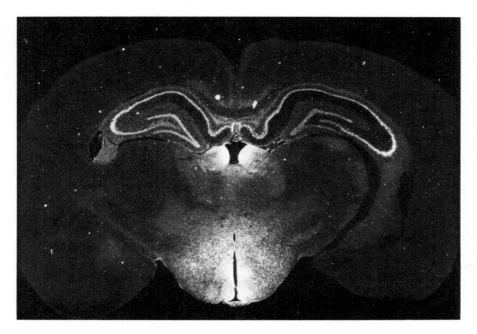

Fig. 2. Transneuronal transfer of ^{3}H-adenosine-derivatives to target neurons of septal fibers in the diencephalon, 2 days after ^{3}H-adenosine injection into septal nuclei. Magnified × 15. Dark-field autoradiography

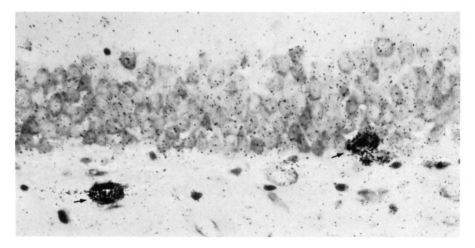

Fig. 3. Transfer of ^{3}H-adenosine derivatives from septal fibers to single neurons in the dentate gyrus *(arrows)* probably representing hippocampal interneurons. Note the non-selective transfer also to perineuronal glial cells. Magnified × 600. [26]

For more detailed studies on the parameters of this transneuronal transfer of adenine compounds we used the afferent projection to the hippocampus which originates in the entorhinal cortex and terminates in well defined layers on the dendrites of hippocampal granule cells. ^{3}H-adenosine was injected into the medial aspect of the entorhinal cortex, and after one day allowed for transport, autoradiographs revealed radioactivity not only in the area where fibers terminate but also in the soma layer of the granule cells [38]. Since the axons of the granule cells terminate exclusively within the hippocampus and do not project back to the site of injection, labeling of granule cells cannot be the result of retrograde transport but must be attributed to a release of labeled adenine compounds from the afferent entorhinal axon terminals and consecutive uptake by the granule cells. The demonstration of such a transneuronal transfer of adenine compounds is not indicative of a purinergic system using ATP as principal transmitter. A release of adenine compounds is also found from septal afferent fibers projecting preferentially to the interneurons of the hippocampus [26] (Fig. 3), and there is convincing evidence that these fibers use acetylcholine as the principal transmitter [21]. Thus, release of adenine compounds seems to accompany transmitter release.

Whether or not such findings reflect, at least to some extent, an intercellular exchange of RNA is a matter of current discussion (see also [12a]). If we analyse the material accumulating within the one day allowed for transport to the hippocampal target area, we find 95%–98% in the TCA-soluble fraction, clearly indicating that the vast majority of the adenine compounds arriving by axonal transport is still in the form of nucleotides and nucleosides. The small fraction of TCA-insoluble material recovered from the terminal area, and probably reflecting RNA, is not indicative of RNA-transport. Its presence can be explained by a local synthesis in the nerve- and glial cells of the target area using transported and released nucleosides as precursors.

In order to get more direct information about the form in which adenine compounds are transported down the axon, we injected ^{3}H-adenosine into the eyeball of rats and analyzed the material recovered from the optic nerve and tract at transport times of 1, 2, and 5 days [5]. The fast moving bulk of radioactivity was found to consist nearly completely of TCA-soluble material, amounting to 96%. In all the segments reached by the fast transported fraction the relative proportion of TCA-insoluble compounds was about the same which is not indicative for a separately moving RNA fraction but may represent glial synthesis. The relative proportion of TCA-insoluble material in the analyzed nerve segments is dramatically increased if uridine is used as precursor instead of adenosine. In this case the TCA-insoluble fraction amounts to 60% of total radioactivity. Since it is unlikely that uridine-labeled RNA is more preferentially transported than adenosine-labeled RNA, the data more likely indicate differences in the release of these two nucleosides and consequently a different availability of these nucleosides for local incorporation into RNA. Further experiments are needed to substantiate this interpretation. If it is correct, one may speculate that RNA metabolism in the nerve cell environment may be influenced to some extent by a differential supply with RNA precursors from the neuronal source.

Dendritic Transport and Release from Dendrites

Adenine compounds are transported with the fast dendritic transport system and reach the finer peripheral branches of the dendritic tree in a motoneuron within minutes after export from the nerve cell soma [32]. This has been shown autoradiographically after intracellular iontophoresis of ^{3}H-adenosine. If 10 min are allowed for transport, the grain density in the peripheral branches is as high as near the cell soma, the site of injection, indicating that the intracellular convection of adenine nucleotides along this route is rather effective. Some of the transported material is extruded through the dendritic membrane and appears incorporated in neighbouring glial cells and cells of the vessel walls (Fig. 4). Such labeled cells are found all around the extension area of the dendritic tree up to some 100 μm distant from the labeled dendritic branches. This cell labeling appears to be rather selective; among the highly

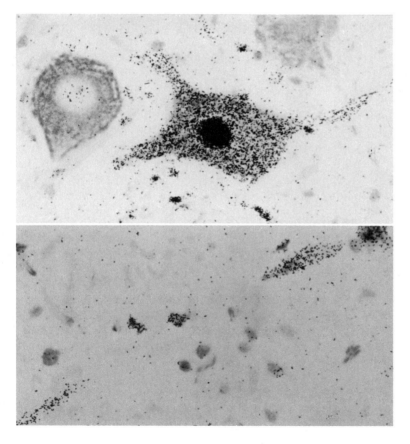

Fig. 4. Release of ^{3}H-adenosine derivatives from dendrites. *Above,* incorporation in the soma; *below,* labeled glial cells near dendritic branches indicating release. Note the presence of unlabeled glial cells. Magnified × 600. Autoradiographs, counterstained with toluidine-blue

labeled glial cells there are always cells which do not show any silver grains. It appears unlikely that this reflects such marked differences in RNA metabolism. A more likely explanation is that some glial cells have access to the released adenine compounds and others are not able to take them up.

The Intracerebral Channelling of Adenine Compounds and its Controlling Factors

All the pathways outlined above by which the exchange of material between the neuron and its environment may occur are used by adenine compounds; there is a release from the dendritic tree as well as from the axon and the axon terminal. From the extracellular space the released adenine compounds are usually taken up again into cells, i.e. glial cells, cells of the vessel wall or nerve cells. In the last case the adenine compounds may again be translocated over longer distances. If the site of re-uptake is not the nerve cell soma, with a high demand for nucleotides, but the axon terminal. then the adenine compounds taken up are effectively transported retrogradely in the axon [12]. Concerning the efficiency of retrograde transport, there again seem to be marked differences for the different nucleosides; adenine compounds are reported to be more efficiently transported retrogradely than uridine compounds [43].

This intracerebral channelling of nucleoside derivatives via intraneuronal transport, release, re-uptake and transport is not uncontrolled but can be regulated at the sites where cell boundaries have to be crossed. First, the release of adenosine derivatives from axon terminals is related to nerve cell activity and increases when the afferent fibers are activated. This has been shown both in vivo and in vitro experiments. In the latter a selective loading of entorhinal afferents to hippocampus neurons was achieved by using axonal transport of radioactive adenosine derivatives as vehicle. After one day allowed for transport hippocampus slices were prepared allowing the selective stimulation of various afferent fiber systems and the recovery of released compounds from the superfused bathing fluid. Only the stimulation of the in vivo labeled pathway led to a significant increase in the release of radioactive material as measured by scintillation counting [19]. In a similar in vivo experiment synaptic activation of radioactively labeled entorhinal fibers by suprathreshold pulses at 5 Hz were found to increase significantly the transfer of radioactive material to the postsynaptic compartment [35]. In contrast to this activity-related release from axon terminals the release of adenine compounds from the axonal shaft seems to be unaffected by functional activation as recently described for the optic nerve of fish (see Grafstein and Edwards, this volume). Whether or not the release from dendrites is activity-related so far remains an open question. It is also not known to what extent the release along the different routes contributes to the total release of adenine compounds from neurons. Nevertheless, nerve cell activity can be regarded as a powerful factor by which release is controlled. This is clearly demonstrated by the experiments of McIlwain who found a severalfold increase in the release of radioactive nucleotides and nucleosides upon electrical field stimulation from brain slices previously loaded with ^{3}H-adenosine [24].

Intracerebral channelling of adenine compounds seems to be further specified by their cellular re-uptake from the extracellular space. There is some evidence, e.g. from the abovementioned in vitro experiments of McIlwain, that the adenine compounds are released at least partly in the form of nucleotides, predominantly AMP (e.g. [3]). Since nucleotides are unable to cross membranes by free diffusion, they have to be

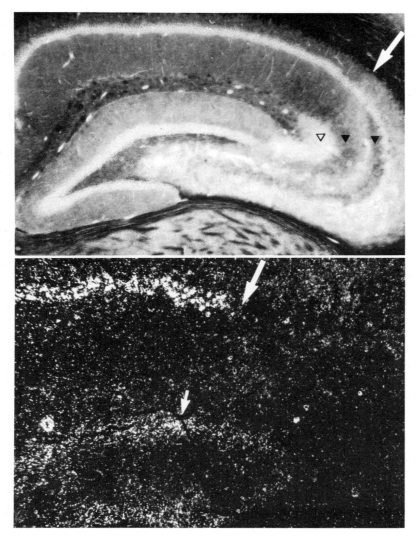

Fig. 5. *Above:* Selective distribution of 5'-nucleotidase in the hippocampus (histochemical demonstration). Afferent entorhinal fibers to CA$_1$ neurons *(left from arrow)* terminate in an area of high enzyme activity, those to CA$_3$ *(right from arrow)* in an area with no demonstrable activity ($\triangledown$). *Below:* After entorhinal injection transfer is seen to CA$_1$ neurons *(left from arrow)*; notice the densely labeled nerve cell somas), but not to CA$_3$-neurons *(right from arrow)*. Magnified × 120. [28]

brought into a membrane permeable form and broken down to adenosine in order to allow their cellular re-uptake. The enzyme hydrolyzing AMP is 5′-nucleotidase which shows a selective distribution in the brain as verified histochemically. In the hippocampus of rat, for instance, a high enzyme activity is found diffusely distributed in the molecular layer of regio superior containing the CA_1 neurons but only in one certain lamina of the molecular layer of regio inferior (Fig. 5). Thus, entorhinal fibers terminating on CA_1 neurons do so within an area of high 5′-nucleotidase activity whereas those terminating on CA_3 neurons make their connections in an area where histochemically no 5′-nucleotidase is demonstrable. If the transfer of ^{3}H-adenosine derivatives from entorhinal fibers to CA_1 and CA_3 neurons was tested autoradiographically, we found the soma layer of CA_1 neurons densely labeled although no activity was seen over the somas of CA_3 neurons [28] (Fig. 5). One possible explanation may be that the lack of transfer to CA_3 is due to the apparent absence of 5′-nucleotidase bringing the released nucleotides into a membrane-permeable form. Therefore, the selective distribution of 5′-nucleotidase should be considered as another possible means of specifying the intracerebral channelling and the local level of extracellularly available adenine compounds.

Electron microscopical histochemical studies have shown that glial cells are the main carriers for the ectoenzyme 5′-nucleotidase [16]. Enzyme activity was seen often associated with the plasma membranes of astrocytic processes engulfing synaptic complexes and also with oligodendroglial cells and microglial cells. A dramatic proliferation of the latter is often found under pathological conditions, e.g. accompanying the primary reaction of motoneurons after nerve transection. As seen e.g. in the regenerating facial nucleus, this event is associated with a drastic increase in 5′-nucleotidase [15]. Such a locally restricted increase in 5′-nucleotidase activity is also found in the terminal area of fibers which undergo degeneration. Thus, after removal of the entorhinal cortex, a distinct band of high 5′-nucleotidase activity is found in the molecular layer of the dentate gyrus in the hippocampus, exactly where these fibers terminate. On the basis of such observations one may speculate that one function of reactive glial cell proliferation is to increase the facilities for the extracellular production of adenosine in a certain area. In view of the anticipated action of adenosine as a neuronal signal this might be one of several mechanisms by which cell functions are adjusted to the functional demand.

Action of Adenosine as Modulatory Neuronal Signal

Adenosine and AMP in micromolar concentrations were found to depress extracellularly recorded evoked potentials from both somatic and dendritic layers in the hippocampus slice [8, 37]. Current source density analysis and intracellular recordings suggest that the primary action of these substances is an influence on synaptic transmission [37]. Moreover, inhibitory as well as excitatory circuitry appear to be subject to adenosine modulation [18]. It thus appears that adenosine action not only leads to an overall quantitative change of synaptic activation but also to a qualitative change in the evoked activity pattern. In as much as the discharge of hippocampal neurons is characterized by a balance of inhibitory and excitatory influences, adenosine and its nucleotides could play a critical role in modulating the output of this system. This modulatory effect on synaptic transmission appears to be mediated via a high affinity A1-receptor [25].

Another effect of adenosine, well-known to biochemists since several years and not related to the electrophysiologically measured effect, is the drastic increase in cAMP synthesis in the brain [27]. In this respect adenosine acts together with other monoamine transmitters.

A third interesting effect of adenosine is its action on the cerebral vasculature where a pronounced dilatation of the vessel walls is elicited. One may therefore speculate that adenosine released from neurons is used as a neuronal signal in order to control local cerebral blood flow and to adjust it to the functional demand. That such a correlation between nerve cell activity and local blood flow exists is known e.g. from clinical studies of Ingvar who found a highly localized increase in blood flow restricted to those brain regions which were activated [13]. Adenosine has also been discussed as being involved in the generation of the complex clinical syndrome called migraine. Here, adenosine-induced dilatation and a direct interaction with pain sensitive fibers is discussed [4].

Several other actions of adenosine are currently under discussion, e.g. its involvement in mediating the effect of opiates, but need further elucidation. But that adenosine acts in a more or less broad sense as a neuronal signal influencing various cell functions seems to be well established. The intra- and transneuronal transport routes and the parameters by which relase and re-uptake is controlled are certainly most critical points by which its functional effects are determined.

References

1. Alvarez J, Püschel M (1972) Transfer of material from efferent axons to sensory epithelium in the goldfish vestibular system. Brain Res 37:265−278
2. Austin L, Bray JJ, Young RJ (1966) Transport of proteins and ribonucleic acid along nerve axons. J Neurochem 13:1267−1269
3. Barberis C, McIlwain H (1976) 5′-adenine mononucleotides in synaptosomal preparations from guinea pigs neocortex: their change on incubation, superfusion and stimulation. J Neurochem 26:1015−1021
4. Burnstock G (1981) Pathophysiology of migraine: a new hypothesis. Lancet June:1397−1398
5. Czlonkowska A, Schubert P, Kreutzberg GW (1982) (in preparation)
6. Droz B, Leblond CP (1963) Axonal migration of proteins in the central nervous system and peripheral nerves as shown by autoradiography. J Comp Neurol 121:325−345
7. Droz B, Koenig HL, Di Giamberardino L (1973) Axonal migration of protein and glycoprotein to nerve endings. I. Radioautographic analysis of the renewal of protein in nerve endings of chicken ciliary ganglion after intracerebral injection of ^{3}H-lysine. Brain Res 60:93−127
8. Dunwiddie TV, Hoffer BJ (1980) Adenine nucleotides and synaptic transmission in the in vitro rat hippocampus. Br J Pharmacol 69:59−68
9. Globus A, Lux HD, Schubert P (1968) Somadendritic spread of intracellularly injected glycine in cat spinal motoneurons. Brain Res 11:440−445

10. Grafstein B (1971) Transneuronal transfer of radioactivity in the central nervous system. Science 172:177–179
11. Grafstein B, Laureno R (1973) Transport of radioactivity from eye to visual cortex in the mouse. Exp Neurol 39:44–57
12. Hunt SP, Künzle H (1976) Bidirectional movement of label and transneuronal transport phenomena after injection of [³H]adenosine in the central nervous system. Brain Res 112: 127–132
12a. Ingoglia NA, Zanakis MF (1982) Axonal transport of 4S RNA. In: Weiss DG (ed) Axoplasmic transport. Springer, Berlin Heidelberg New York, pp 161–169
13. Ingvar DH (1973) Cerebral blood flow metabolism in complete apallic syndroms, in states of severe dementia and akinetic mutism. Acta Neurol Scand 49:233–244
14. Korr IM, Wilkinson PN, Chornock FW (1967) Axonal delivery of neuroplasmic components to muscle cells. Science 155:342–345
15. Kreutzberg GW, Barron KD (1978) 5'-Nucleotidase of microglial cells in the facial nucleus during axonal reaction. J Neurocytol 7:601–610
16. Kreutzberg GW, Barron KD, Schubert P (1978) Cytochemical localization of 5'-nucleotidase in glial plasma membranes. Brain Res 158:247–257
17. Kreutzberg GW, Tóth L, Kaiya H (1975) Acetylcholinesterase as a marker for dendritic transport and dendritic secretion. Adv Neurol 12:269–281
18. Lee KS, Schubert P (1982) Modulation of an inhibitory circuit by adenosine and AMP in the hippocampus. Brain Res (in press)
19. Lee KS, Schubert P, Gribkoff V, Brimmer S, Lynch G (1982) A combined in vivo/in vitro study of the presynaptic release of adenosine derivatives in the hippocampus. J Neurochem 38:80–83
20. Lux HD, Schubert P, Kreutzberg GW, Globus A (1970) Excitation and axonal flow. Autoradiographic study on motoneurons intracellularly injected with ³H-amino acid. Exp Brain Res 10:197–204
21. Lynch G, Matthews DA, Mosko S, Parks T, Cotman CW (1972) Induced AChE-rich layer in rat dentate gyrus following entorhinal lesions. Brain Res 42:311–318
22. Miani N (1971) Transport of S-100 protein in mammalian nerve fibers and transneuronal signals. Acta Neuropathol Suppl 5:104–108
23. Nieoullon A, Chéramy A, Glowinski J (1977) Release of dopamine in vivo from cat substantia nigra. Nature (London) 266:375–377
24. Pull I, McIlwain H (1977) Adenine mononucleotides and their metabolites liberated from and applied to isolated tissues of the mammalian brain. Neurochem Res 2:203–216
25. Reddington M, Lee KS, Schubert P (1982) Depression of evoked potentials in a hippocampal slice preparation by adenosine: mediation via an A1-adenosine receptor characterized by a ³H-cyclohexyladenosine binding assay. Neurosci Lett 28:275–279
26. Rose G, Schubert P (1977) Release and transfer of [³H]adenosine derivatives in the cholinergic septal system. Brain Res 121:353–357
27. Sattin A, Rall TW (1970) Cyclic AMP content of guinea pig cerebral cortex slices. Mol Pharmacol 6:13–23
28. Schubert P, Komp W, Kreutzberg GW (1979) Correlation of 5'-nucleotidase activity and selective transneuronal transfer of adenosine in the hippocampus. Brain Res 168:419–424
29. Schubert P, Kreutzberg GW (1974) Axonal transport of adenosine and uridine derivatives and transfer to postsynaptic neurons. Brain Res 76:526–530
30. Schubert P, Kreutzberg GW (1975) Parameters of dendritic transport. Adv Neurol 12:255–268
31. Schubert P, Kreutzberg GW (1975) [³H]adenosine, a tracer for neuronal connectivity. Brain Res 85:317–319
32. Schubert P, Kreutzberg GW (1975) Dendritic and axonal transport of nucleoside derivatives in single motoneurons and release from dendrites. Brain Res 90:319–323
33. Schubert P, Kreutzberg GW (1976) Communication between the neuron and the vessels. In: Cervós-Navarro et al. (eds) The cerebral vessel wall. Raven Press, New York, pp 207–213
34. Schubert P, Kreutzberg GW, Lux HD (1972) Neuroplasmic transport in dendrites: effect of colchicine on morphology and physiology of motoneurons in the cat. Brain Res 47:331–343

35. Schubert P, Lee K, West M, Deadwyler S, Lynch G (1976) Stimulation-dependent release of ^{3}H-adenosine derivatives from central axon terminals to target neurones. Nature (London) 260:541–542
36. Schubert P, Lux HD, Kreutzberg GW (1971) Single cell isotope injection technique, a tool for studying axonal and dendritic transport. Acta Neuropathol 5:179–186
37. Schubert P, Mitzdorf U (1979) Electrophysiological evaluation of the depressive effect of adenosine on evoked potentials in hippocampus slices. Brain Res 172:186–190
38. Schubert P, Rose G, Lee K, Lynch G, Kreutzberg GW (1977) Axonal release and transfer of nucleoside derivatives in the entorhinal-hippocampal system: an autoradiographic study. Brain Res 134:347–352
39. Schwab M, Thoenen H (1977) Selective transsynaptic migration of tetanus toxin after retrograde axonal transport in peripheral sympathetic nerves: a comparison with nerve growth factor. Brain Res 122:459–474
40. Specht S, Grafstein B (1973) Accumulation of radioactive protein in mouse cerebral cortex after injection of ^{3}H-fucose into the eye. Exp Neurol 41:705–722
41. Specht SC, Grafstein B (1977) Axonal transport and transneuronal transfer in mouse visual system following injection of [^{3}H]fucose into the eye. Exp Neurol 54:352–368
42. Wise SP, Jones EG (1976) Transneuronal or retrograde transport of [^{3}H]adenosine in the rat somatic sensory system. Brain Res 107:127–131
43. Wise SP, Jones EG, Berman N (1978) Direction and specificity of the axonal and transcellular transport of nucleosides. Brain Res 139:197–217

Axonal Transport, Neurosecretory Vesicles, and the Endocrine Neuron

HAROLD GAINER [1], JAMES T. RUSSELL [1], and MICHAEL J. BROWNSTEIN [2]

Introduction

As its name implies, the endocrine neuron combines the attributes of both neurons and endocrine cells. While it possesses the morphological and electrophysiological characteristics of neurons, it is also specialized to biosynthesize, axonally transport, and secrete large quantities of peptide hormones. In as much as it is a bona fide neuron, one would expect that axonal transport plays the same roles in endocrine neurons as it does in other neurons. Indeed, our studies of the multiple components of axonal transport in the endocrine neurons of the hypothalamo-neurohypophysial system [7] are in agreement with the classifications of axonal transport components presented by others in this volume [2, 3].

What makes the endocrine neuron distinctive, is its high level of committment to secretion, and therefore, to the large production and axonal transport of hormone-containing neurosecretory vesicles [18, 21]. These 160—190 nm neurosecretory vesicles (in mammalian neurosecretory systems) are transported at a rate of about 150 mm/day; and contain the hormones oxytocin, vasopressin, and their respective neurophysins (the 10,000 M.W. binding proteins [18, 19, 22, 23]). Each hormone and its associated neurophysin is located in a separate neuron in the form of intragranular complexes in the vesicles, with an association constant of about 10^5 and with a maximum binding at pH 5.5 [18, 23].

The neurosecretory vesicle represents one of the few organelles which can be clearly associated with the fast axonal transport component. It is our contention that this organelle is more than simply a vehicle in transport and secretion processes. It also appears to play an active role in the biosynthesis of the hormones and the neurophysins.

---

1 Section on Functional Neurochemistry, Laboratory of Developmental Neurobiology NIH, NICHD, Bldg. 36, Bethesda, MD 20205, USA
2 Laboratory of Clinical Science, National Institute of Mental Health, Bethesda, MD 20205, USA

Axoplasmic Transport in Physiology and Pathology
(ed. by D.G. Weiss and A. Gorio)
© Springer-Verlag Berlin Heidelberg 1982

Biosynthesis of the Neurohypophysial Peptides and Neurophysins

As has been found for many other peptide hormones [16, 29], the neurohypophysial peptide hormones are first synthesized as larger precursor proteins (prohormones), which are subsequently proteolytically cleaved into smaller biologically active peptides and proteins. Since we have recently reviewed the status of this field [4], only a brief summary will be presented here.

In 1964 Sachs and Takabatake [28] hypothesized that vasopressin was biosynthesized via a prohormone mechanism, and later proposed that the precursor contained both vasopressin and its associated neurophysin [27]. More recent pulse-chase studies [9, 11, 12, 25] have led to the identification of two precursors in the rat, one for oxytocin (OT) and its neurophysin (Np), and the other for arginine vasopressin (AVP) and its neurophysin. The current status of information about these prohormones is given in Table 1. The vasopressin-neurophysin precursor (pro-pressophysin) is a protein of about 19,500 M.W. with an isoelectric point (pI) equal to 6.1, and contains vasopressin, the vasopressin-associated neurophysin (pI = 4.8), and glycopeotide of about 8,000 M.W. In contrast, the oxytocin-neurophysin precursor (pro-oxyphysin) does not contain carbohydrate (CHO), is smaller (M.W. about 15,000), has a pI = 5.4, and contains oxytocin, the oxytocin-associated neurophysin (pI = 4.6), and possibly another cysteine-containing peptide.

Table 1. Properties of rat nonapeptide/neurophysin precursors

Molecule	pI	Molecular weight		Molecule contains [a]		
		15% SDS gel [a]	G75-GuHCL [b]	NP	AVP or OT	CHO
Pro-pressophysin (Pro-PP)						
Precursor	6.1	19,500	20,500	+	+	+
Intermediate	5.6	18,000	−	+	+	+
Pro-Oxyphysin (Pro-OP)						
Precursor	5.4	15,000	18,700	+	+	−
Intermediate	5.1	14,000	−	+	+	−
Neurophysin (Np)						
Neurophysin (AVP)	4.8	10–12,000	10–12,000	+	−	−
Neurophysin (OT)	10–12,000		10–12,000	+	−	−

[a] M.W. determined by electrophoresis of purified precursors, intermediates, and neurophysins (using perparative isoelectric focusing) on one dimensional 15% SDS gels

[b] M.W. determined by chromatography on G75 Sephadex columns in the presence of 6 M guanidinium-HCl

[c] Presence of Np in isolated proteins was determined by direct immunoprecipitation with antirat Np, or by demonstrating binding of 10K tryptic product to LVP-affinity columns. Presence of AVP or OT was determined by binding of tryptic peptide product to Np-affinity columns. and by immunoprecipitation of tryptic peptide antobides to AVP (for Pro-PP) or OT (for Pro-OP). Presence of carbohydrate (CHO) on protein determined by binding to CON-A affinity columns

Recent studies on cell-free translation of bovine hypothalamic m-RNA in wheat germ extracts and rabbit reticulocyte lysates (see Richter et al. in [16]) are in good agreement with the data shown in Table 1 for rat precursors. In conventional in vitro translation systems the precursors are synthesized as pre-prohormones. The molecular weights of bovine pre-pro-pressophysin and pre-pro-oxyphysin are 21,000 and 16,500, respectively, when the translation is performed in the presence of microsomal membranes, the signal sequences are removed and the bovine pro-pressophysin and pro-oxyphysin molecular weights are 19,000 and 15,500, respectively. As in the case of the rat prohormones only the bovine-pro-pressophysin can be glycosylated. No pI values for the in vitro translated prohormones have been reported.

The Neurosecretory Vesicle: A Role in the Post-Translational Processing of Precursors

Based on indirect evidence, Sachs and his colleagues [27] proposed that the varopressin prohormone was converted after packaging into neurosecretory vesicles. More recent experiments [4, 9, 11, 12] have shown that the prohormones are axonally transported (see Fig. 1) and that conversion of the prohormones occurs during axonal transport in the axons and terminals. In recent experiments performed in our laboratory, supraoptic nuclei (SON) of rats were pulsed with ^{35}S-cysteine for 20 min in vivo, and then the hypothalami (included cell bodies and axons of SON neurons) were isolated, homogenized in isotonic sucrose and sedimented by centrifugation.

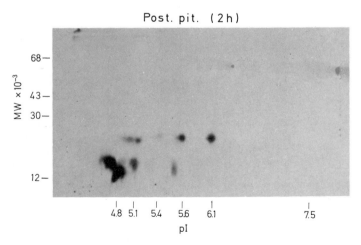

Fig. 1. Autoradiograph of a 2-dimensional gel containing ^{35}S-cysteine labeled proteins which were obtained from rat posterior pituitary 2 h after injection of the labeled amino acid (^{35}S-cysteine) into the SON. Note, in addition to labeled neurophysin (around pI 4.8), prominent labeled spots corresponding to the pI 6.1 pro-pressophysin and pI 5.6 intermediate, and the pI 5.1 pro-oxyphysin intermediate

When this sedimented fraction (containing neurosecretory vesicles and other cell organelles) was incubated in vitro, none of the labelled prohormones in the SON were
converted to neurophysins. However, in a similar experiment, but one which allowed
for a 60—90 min chase period in vivo following the pulse label, it was possible to
demonstrate conversion of labelled prohormone to neurophysins in the in vitro incu-
bations. This implied that a translocation step(s) in vivo was required before conversion could occur. Since we know that the prohormones can be transported to the
axons of the median eminence and neurohypophysis by 60—90 min [11, 12, 21],
it is likely that translocation and packaging of the prohormones into neurosecretory
vesicles occurred by this time. We believe that these data, taken as a whole, provide
a compelling case for the neurosecretory vesicle as the principal site of conversion of
these prohormones.

Given the hypothesis that the neurosecretory vesicle is the major site for prohormone conversion, then one would predict that appropriate converting-enzymes should
be present in the vesicles. Inspection of the amino acid sequences of all those prohormones for which such information is available (see [29]), reveals that they all contain pairs of basic amino acid residues (i.e., lysine and/or arginine) at their expected
sites of enzymatic cleavage. Hence it has been proposed [29] that the converting-
enzymes for prohormones should involve trypsin-like endopeptidases and carboxypeptidase-B-like enzymes. Recently, Fletcher et al. [8] have described a tryptic-like
enzymatic activity in secretory granules from anglerfish pancreatic islet tissue which
successfully converts prosomatostatin. The enzyme appears to be vesicle associated
thiol proteinase, with a possible specificity for arginine residues. Thie enzymatic
activity has a pH optimum of around 5.2, but is not affected by known inhibitors of
cathepsin B.

Several indirect observations suggest an acidic (pH 5—6) milieu within the neurosecretory vesicle. First, vasopressin and oxytocin which are believed to be bound to
their respective neurophysin carrier proteins in the vesicle, bind at an optimum pH
of 5.5 [23]. Second, morphological studies by Morris et al. [18] indicate that the
dense cores of neurosecretory vesicles in the posterior pituitary are most stable if
fixation is performed at acidic pHs (i.e., around pH 5). This issue of intravesicular pH
is a central one since it defines at least one aspect of the microenvironment within
the vesicles in which the converting enzymes are presumed to work. Neurosecretory
vesicles isolated using iso-osmolar density gradients [24] are highly stable, and thus,
allow for the measurement of pH gradients across the vesicle membrane under a
variety of experimental conditions. Measurements of such pH gradients (and therefore, or intravesicular pH) have been made in our laboratory using the ^{3}H-methyl-
amine distribution method [26]. The results indicate that the pH of the intravesicular
space is around 5.5. Given this data, one would expect that if intravesicular converting
enzymes do exist, then they should have pH optima in the pH 5—6 range.

Using labelled prohormones as substrates we have been able to detect trypsin-like
and carboxypeptidase-B-like activites in lysates of the purified neurosecretory vesicles
(Loh, Chang, Hook, Russell, Brownstein, and Gainer, unpublished data). These converting-enzymes have pH optima around pH 5.0, and are either inactive or poorly
active at higher pH values (i.e., at pH 7.4 or greater). At present, we are attempting
to characterize the specificities of these enzymes. What is clear at present, is that the

intravesicular tryptic-like enzyme does not cleave substrates with only one basic amino acid residue (as are used in conventional trypsin assays), but appears to require pairs of basic residues as are generally found in prohormones.

Conclusions

In this paper, we have focused on the neurosecretory vesicle as the principal vehicle for the fast axonal transport of peptides and proteins in the hypothalamo-neurohypophysial system. In addition, we have presented some evidence in support of the notion that another biologically significant event is occurring within this vesicle (i.e., the enzymatic processes associated with prohormone conversion). Whether related processes occur in other anterogradely transported vehicles in other peptidergic neurons, as well as "conventional" neurons, is an open question.

Very little is known about the nature of the vehicle for fast axonal transport in conventional neurons, except that it is a membrane-bound organelle derived from the Golgi complex. In this regard it appears analogous to secretory vesicles in peptidergic systems [10, 13, 14, 20]. However, in the axons of most conventional neurons no organelles analogous to the secretory vesicles are apparent. Instead, the axonal vehicle for fast transport has been alternatively identified as the complex agranular endoplasmic reticulum (AER) in axons [6] or as tubule-like vesicular elements (see [15]). The compromise position is that the tubular elements represent the actual vehicles for movement, and that these elements can transiently fuse with and be recovered from the axonal AER system. At present, there is no data available for or against this view. At first glance, this problem might seem irrelevant for endocrine neurons, since neurosecretory vesicles are so prominent in their axons. However, several investigators have argued from morphological data [1, 5, 17] that the axonal AER may also be a vehicle for hormone transport in endocrine neurons especially during increased secretion (a polemic against this view can be found in [18]). In view of the important role played by intravesicular acidic converting enzymes in the production of completed peptide products for secretion (discussed above), it is difficult to envision what selective advantage an AER transport system would have over neurosecretory vesicles during periods of enhanced secretion [1, 17]. Further characterization and isolation of the converting enzymes may provide an opportunity to resolve this controversy. Given an antibody to these enzymes, it should be possible, by immunocytochemical techniques, to determine whether they are present in the AER during periods of increased secretory activity.

References

1. Alonso G, Assenmacher I (1979) The smooth endoplasmic reticulum in neurohypophysial axons of the rat: possible involvement in transport, storages, and release of neurosecretory material. Cell Tissue Res 199:415–429

2. Baitinger C, Levine J, Simon C, Skene P, Willard M (1982) Characteristics of axonally transported proteins. In: Weiss DG (ed) Axoplasmic transport. Springer, Berlin Heidelberg New York, pp 110–120

3. Brady ST, Lasek RJ (1982) The slow components of axonal transport: movements, compositions and organization. In: Weiss DG (ed) Axoplasmic transport. Springer, Berlin Heidelberg New York, pp 206–217

4. Brownstein MJ, Russell JT, Gainer H (1980) Biosynthesis, axonal transport, and release of posterior pituitary hormones. Science 207:373–378

5. Castel M, Dellmann HD (1980) Thiamine pyrophosphatase activity in the axonal smooth endoplasmic reticulum of neurosecretory neurons. Cell Tissue Res 210:205–221

6. Droz B, Rambourg A, Koenig HL (1975) The smooth endoplasmic reticulum: structure and role in the renewal of axonal membrane and synaptic vesicles by fast axonal transport. Brain Res 93:1–13

7. Fink DJ, Russell JT, Brownstein MJ, Baumgold J, Gainer H (1981) Multiple rate components of axonally transported proteins in the hypothalamo-neurohypophysial system of the rat. J Neurobiol (in press)

8. Fletcher DJ, Noe BD, Bauer GE, Quigley JP (1980) Characterization of the conversion of a somatostatin precursor to somatostatin by islet secretory granules. Diabetes 29:593–599

9. Gainer H, Brownstein MJ (1978) Identification of the precursors to rat neurophysins. In: Vincent JD, Kordon C (eds) Cell biology of hypothalamic neurosecretion. CNRS Press, Paris, pp 525–542

10. Gainer H, Russell JT, Fink DJ (1979) The hypothalamo-neurohypophysial system: a cell biological model for peptidergic neurons. In: Gotto AM (ed) Brain peptides. Elsevier/North Holland Biomedical Press, Amsterdam New York, pp 139–159

11. Gainer H, Sarne Y, Brownstein MJ (1977) Neurophysin biosynthesis: conversion of a putative precursor during axonal transport. Science 195:1354–1356

12. Gainer H, Sarne Y, Brownstein MJ (1977) Biosynthesis and axonal transport of rat neurohypophysial proteins and peptides. J Cell Biol 73:366–381

13. Hammerschlag R, Stone GC (1982) Fast axonal transport as endomembrane flow. In: Weiss DG (ed) Axoplasmic transport. Springer, Berlin Heidelberg New York, pp 406–413

14. Hökfelt T, Johansson O, Ljungdahl A, Lundberg JM, Schultzberg M (1980) Peptidergic neurones. Nature (London) 284:515–521

15. Ishikawa H, Tsukita S (1982) Morphological and functional correlates of axoplasmic transport. In: Weiss DG (ed) Axoplasmic transport. Springer, Berlin Heidelberg New York

16. Koch G, Richter D (1980) Biosynthesis, modification, and processing of cellular and viral polyproteins. Academic Press, London New York

17. Krisch B (1979) Indication of a granule-free form of vasopressin in immobilization-stressed rats. Cell Tissue Res 197:95–104

18. Morris JF, Nordmann JJ, Dyball REJ (1978) Structure-function correlation in mammalian neurosecretion. Int Rev Exp Pathol 18:1–95

19. North WG, Mitchell TI (1981) Evolution of neurophysin proteins: partial amino acid sequences of rat neurophysins. FEBS Lett 126:41–44

20. Palade G (1975) Intracellular aspects of the process of protein synthesis. Science 189:347–358

21. Pickering BT (1978) The neurosecretory neurone: a model for the study of secretion. Essays Biochem 14:45–81

22. Pickering BT (1982) Effects of colchicine in the hypothalamo-neurohypophysial system of the rat. In: Weiss DG (ed) Axoplasmic transport. Springer, Berlin Heidelberg New York, pp 139–143

23. Pickering BT, Jones CW (1978) The neurophysins. In: Li CH (ed) Hormonal proteins and peptides. Academic Press, London New York, pp 103–158
24. Russell JT (1981) The isolation of purified neurosecretory vesicles from bovine neurohypophysis using iso-osmolar density gradients. Anal Biochem (in press)
25. Russell JT, Brownstein MJ, Gainer H (1980) Biosynthesis of vasopressin, oxytocin, and neurophysins: isolation and characterization of two common precursors (prooxyphysin and propressophysin). Endocrinology 107:1880–1891
26. Russell JT, Holz RW (1981) Measurement of ΔpH and membrane potential in isolated neurosecretory vesicles from bovine neurohypophyses. J Biol Chem 256:5950–5953
27. Sachs H, Fawcett P, Takabatake Y, Portanova R (1969) Biosynthesis and release of vasopressin and neurophysin. Recent Prog Horm Res 25:447–491
28. Sachs H, Takabatake Y (1964) Evidence for a precursor in vasopressin biosynthesis. Endocrinology 75:943–948
29. Zimmerman M, Mumford RA, Steiner DF (eds) (1980) Precursor processing in the biosynthesis of proteins. Ann NY Acad Sci 343

Section 2   The Role of Axoplasmic Transport in Growth and Regeneration

Axonal Transport in the Sprouting Neuron: Transfer of Newly Synthesized Membrane Components to the Cell Surface

KARL H. PFENNINGER [1]

Introduction

During the formation of axons and dendrites, massive increase in cell surface area necessitates rapid expansion of the neuron's plasma membrane. From the diameter of the processes and the rate of advancement of the nerve growth cones, the rate of addition of new membrane to the cell surface can be calculated. In a sympathetic neuron of the rat, grown in culture under the influence of nerve growth factor, elongation of neurites occurs at a rate of approximately 1 mm/day. Their diameter being approximately 0.5 μm, the rate of increase in plasmalemmal surface is approximately 1 μm^2/min per neurite. Thus, at least 1 μm^2 of plasmalemmal precursor material needs to be transported each minute to the surface of each sprouting neurite. A priori, membrane components could be transferred from sites of synthesis to the cell surface singly, in small aggregates, or in already assembled, membranous form; they could be inserted into the plasmalemma at specific sites such as the tip or the base of the sprouting neurite or, alternatively, they could be inserted in random fashion in proximal and distal parts of the neuron. Both Hughes [3] and Bray [1, 2] have suggested that newly synthesized membrane components are inserted into the plasmalemma at the tip of the growing neurite. The analyses carried out in this laboratory confirm this hypothesis, albeit with certain modifications. It follows that axonal transport plays an important role in the transfer of membrane components from sites of synthesis (which are predominantly, if not exclusively, located in the perikaryon) to the surface of the sprouting neuron.

This chapter aims at (a) demonstrating that newly synthesized membrane components are indeed rapidly and selectively transferred to the sprouting neurite and (b) identifying the plasmalemmal precursor on its way from the perikaryon to the distal axolemma. Furthermore, various properties of growing axolemma and its precursor are described. Finally, the implications of our data on pathways and modes of transport of newly synthesized membrane components are discussed.

---

1 Dept. of Anatomy and Cell Biology, College of Physicians and Surgeons of Columbia University, 630 West 168th Street, New York, NY 10032, USA

Axoplasmic Transport in Physiology and Pathology
(ed. by D.G. Weiss and A. Gorio)
© Springer-Verlag Berlin Heidelberg 1982

Transfer of Newly Synthesized Phospholipid into the Growing Neurite

In order to study the fate of newly synthesized phospholipid in the sprouting neuron, explant cultures of rat superior cervical ganglion were grown in vitro until a broad halo of neuritic outgrowth was formed. Such cultures were then pulsed with tritiated glycerol, a precursor of phospholipid, for 15 min and subsequently chased with an excess of non-radioactive glycerol for various periods of time. At the end of the chase, explants containing the neuronal perikarya and neuritic outgrowth were collected separately, extracted with chloroform-methanol, and radioactivity as well as inorganic phosphate were determined in the phospholipid phase. In the perikarya there is a steep increase in specific radioactivity of phospholipid during the pulse period. This, however, is followed by rapid decline in specific radioactivity in the control experiment [7, 10]. By contrast, specific radioactivity of neuritic phospholipid increases only after a lag period of approximately 60 min. A decline in specific radioactivity similar to that measured in the perikarya cannot be found in the neurites during an observation period of approximately 3 h. This result is consistent with the idea of rapid and selective transfer of newly synthesized phospholipid from the perikaryon into the growing neurite. That this transfer is energy-dependent can be shown by poisoning the cultures with the metabolic inhibitor 2,4-dinitrophenol 5 min following the 15-min ^{3}H-glycerol pulse. This blocks completely the decline in specific radioactivity observed in control perikarya as well as the increase in specific radioactivity of the neurites. A further blocking experiment was carried out with colchicine (2.5 × 10^{-6} M). When this drug is applied specific radioactivity of perikaryal phospholipid reaches approximately control levels, but declines only slightly during the longer chase periods. Conversely, the neurite's phospholipid radioactivity does not increase at all. These data indicate that newly synthesized phospholipid reaches the distal end of the neurite by rapid axonal transport via a cytoplasmic pathway and, probably, in the form of assembled membranous structures.

Identification of the Axolemmal Precursor and the Site of its Insertion into the Plasma Membrane

a) Autoradiographic Studies

In order to identify the organelle which transports newly synthesized phospholipid to the distal end of the growing neurite, phospholipids of neurons growing in culture were again pulse-labeled for 30 min with tritiated glycerol and then chased. (As before, superior cervical ganglion neurons of the rat, grown as explants for 4 days in culture, were used as experimental tissue.) At different times after onset of the pulse, the neurons were processed for electron microscopic radioautography. In parallel experiments it was established that, under our experimental conditions, approximately 99% of the precursor ^{3}H-glycerol is incorporated into phospholipid, and that at least

70% of lipid-incorporated radioactivity is retained in the tissue, even after dehydration and embedding in epoxy resin.

The radioautographs were analyzed by both a modification of the technique described by Williams [19] and by the method developed by Salpeter and Bachmann [15]. Detailed results will be presented in a forthcoming publication [9] and can be summarized as follows [7]: (1) The overall grain counts in the growth cone region are first low (Fig. 1; 15 min after onset of pulse) and then increase gradually (Fig. 2; 30 min) to reach a plateau approximately 60 min after onset of the labeling. The relative labeling density of different growth cone membrane compartments (plasma membrane, a class of large clear vesicles characteristic of growing neurites, agranular reticulum, and other organelles such as mitochondria and lysosomes) was assessed with the Williams method, as a function of time during and after the ^{3}H-glycerol pulse. The large clear vesicles, so-called subplasmalemmal vesicles (SPVs), rapidly (at 30 min after onset of labeling) reach far greater density per unit area of membrane (Fig. 2) than all other membrane compartments, including the growth cone's agranular reticulum. As this transient peak diminishes, the plasma membrane's grain density increases. Agranular reticulum, in contrast, does not appear to participate in these relatively rapid changes, but only slowly increases in grain density.

(2) Analysis of the average distance of silver grains from two membrane compartments, SPVs and the plasma membrane (Salpeter method), reveals an analogous picture: concentration of silver grains on and near SPVs, at 15–30 min following onset of labeling, precedes a shift of silver grain distribution from the intracellular SPV compartment to the plasma membrane. This shift is nearly completed by about 60 min.

Radioautographic studies parallel the biochemical analyses presented above and identify SPVs as the vehicle of newly synthesized phospholipid, i.e., as the precursor of the plasma membrane of the nerve growth cone.

b) Pulse-Chase Studies with Lectins

A different and complementary approach to the analysis of membrane expansion is the study of the insertion of new lectin receptors in the growing neuron's plasmalemma. Lectins, proteins which bind highly selectively to specific carbohydrate residues [4], were purified by affinity chromatography and used as such or conjugated with the electron dense marker ferritin [5]. Ferritin conjugates were purified so that only active lectin, bound to the marker in a 1:1 ratio, was present in the final fraction. These experiments have been reported in detail elsewhere [14], and only three of the key experiments are described here.

First, live sympathetic neurons, grown in vitro, were carefully washed free of medium glycoproteins and then aldehyde fixed. The quenched cultures were subsequently exposed to a saturating concentration of a lectin-ferritin conjugate, e.g., that of wheat germ agglutinin (specific for N-acetyl-glucosamine residues) or ricin I (specific for galactosyl residues). Electron microscopic analysis reveals completely uniform labeling of growth cone plasmalemma in these experiments (Fig. 3). An SPV can occasionally be seen open to the extracellular space; in this case its luminal membrane

Figs. 1 and 2. Electron microscopic autoradiographs of growth cone regions in cultures of rat superior cervical ganglia, pulsed with ^{3}H-glycerol for 30 min. In Fig. 1, 15 min after onset of the pulse, few silver grains are evident. However, in Fig. 2, 30 min after onset of the pulse, silver grains are much more frequent, especially over clusters of SPVs *(arrowheads)*. At later times, silver grains increase further in density and shift to a distribution primarily associated with the plasmalemma. *f* growth cone filopodia. For further description see text. × 19,800; *bar* 1 μm

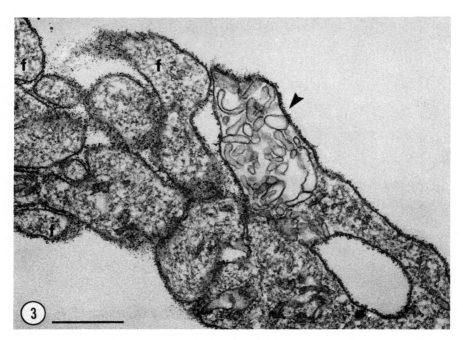

Fig. 3. Labeling with ferritin-conjugated wheat germ agglutinin of part of a nerve growth cone from rat superior cervical ganglion. Prerinsed tissues were first aldehyde-prefixed, quenched with glycine and bovine serum albumin, and then labeled with the lectin. Note uniform labeling of the cell surface including the plasmalemma covering filopodia *(f)* and an SPV cluster *(arrowhead)*. X 37,700; *bar* 0.5 μm. (From [14])

surface is consistently labeled. This suggests that the insides of the vesicles carry lectin receptors similar to those on the cell surface. In the second type of experiment, the carefully washed cultures were labeled live and then allowed to survive in the absence of the label for 3 to 20 min. In this situation, growth cone periphery and, especially, the plasma membrane covering clusters of SPVs are rapidly fred of the label (Fig. 4). This uneven distribution of the marker is strikingly different from the uniform labeling seen in the previous experiment. Internalization of the label into SPVs does not seem to occur. In the third type of experiment, live, growing neurons were exposed to a saturating concentration of the unconjugated lectin, then allowed to survive in the absence of the label, aldehyde fixed, and subsequently (after quenching of aldehyde) re-labeled with the ferritin conjugate of the same lectin. In this experiment, only those lectin receptors which have previously been inaccessible to the lectin carry the ferritin marker, i.e., those which appeared on the cell surface during the chase, or survival, period. The labeling image thus obtained is exactly complementary to the one in the previous experiment (Fig. 5; cf. Fig. 4): the surfaces of growth cone filopodia and, especially, of SPV clusters are now densely labeled while the surrounding plasmalemma carries few ferritin particles. If the same experiment is carried out on ice, selective labeling of vesicle cluster surfaces and filopodia is not observed.

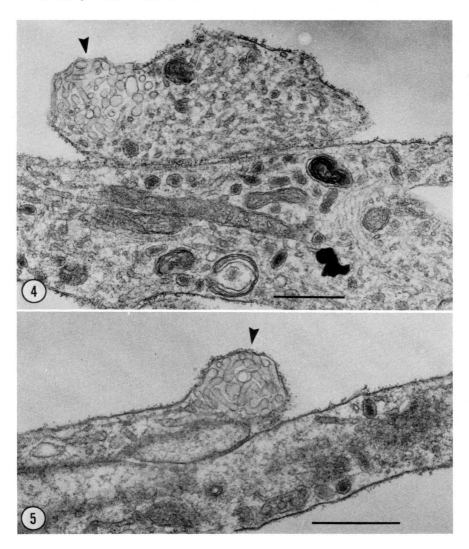

Fig. 4. Pulse-chase experiment on nerve growth cones (superior cervical ganglion of the rat) with ferritin-wheat-germ agglutinin. Prerinsed culture was exposed for 5 min to the lectin conjugate and then chased in the absence of the lectin for 15 min before being processed for electron microscopic analysis. Note dense labeling of the cell surface except for plasmalemma covering SPV cluster *(arrowhead)*. × 37,000; *bar* 0.5 μm

Fig. 5. Double-label pulse-chase experiment with wheat germ agglutinin on preparation similar to that described for Fig. 4. Culture was first exposed to unconjugated lectin, chased in the absence of the ligand for 15 min, then aldehyde-fixed, and finally quenched and relabeled with ferritin-wheat-germ agglutinin. Note selective, much denser labeling of SPV cluster *(arrowhead)* compared to low label density in other plasmalemmal domains. This image is complementary to that seen in Fig. 4. × 46,800; *bar* 0.5 μm. (Figs. 4 and 5 from [14]

These data, together with a variety of control experiments, indicate that the most distal region of the growth cone, specifically the plasma membrane covering clusters of SPVs, is the site of insertion of new lectin receptors during neuritic growth. These results further support the concept that SPVs are the precursor of the plasmalemma.

Cell Surface Polarity of the Growing Neuron

The morphological polarity of the growing neuron raises the question of whether the plasma membrane may exhibit regional specialization as well. Preferential insertion of newly synthesized phospholipid and certain glycoconjugates at the growth cone suggests that growth cone plasmalemma may indeed differ in its composition from that of the more proximal portions of the neuron. This question has been investigated with two methods, quantitative lectin labeling and freeze-fracture.

a) Lectin Binding Studies

The lectin-ferritin conjugates described above enabled us to quantitate in different cellular regions the density per unit area of membrane of certain superficially exposed carbohydrate residues. This was done with a series of different lectins [13]. In brief, the results are that glycoconjugate composition differs significantly in growth cone, neuritic shaft, and perikaryon. While certain lectin receptors are more frequent in the growth cone area than in the perikaryon, others may exhibit the inverse relative distribution. Yet another group of lectin receptors is distributed evenly throughout the plasmalemma of the growing neuron. It is important to note that these relative and absolute densities of lectin binding sites are characteristic for different types of neurons [11, 12, 13]. Our results demonstrate that cellular polarity of the growing neuron is reflected in the topographical distribution of plasmalemmal glycoconjugates.

b) Freeze-Fracture Studies

It has been known for some time that the nerve growth cone, albeit rich in plasma-lemmal lectin binding sites, contains very few plasmalemmal intramembranous particles (IMPs) (Figs. 6 and 7) [6, 8]. This phenomenon was investigated in greater detail in the growing olfactory nerve of the bullfrog, *Rana catesbeiana* [16, 17]. As in the cultured rat neurons, IMPs are very sparse in the distal regions of the growing axon. Along the axonal shaft, overall particle density forms an exponentially decreasing gradient in proximo-distal direction. Particle composition, i.e., frequency of IMPs of specific size, also varies as a function of the distance from the perikaryon. Again, the densities of specific sizes of IMPs obey distally decreasing exponential functions; however, gradients are shallower for the smaller IMPs and steeper for the larger ones. Such a particle distribution parallels a family of size-dependent diffusion gradients.

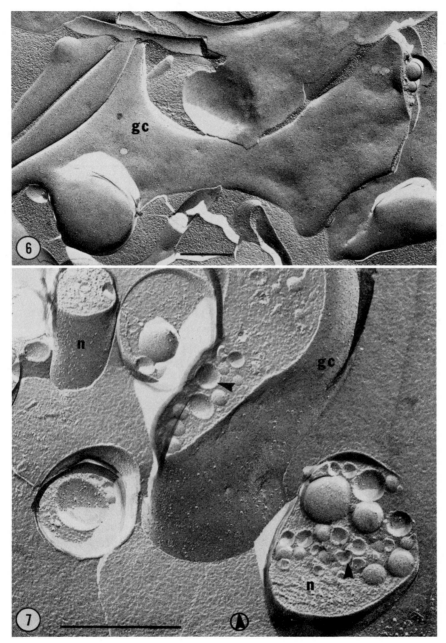

Figs. 6 and 7. Freeze-fracture electron micrographs of growth cone (*gc* protoplasmic leaflet) in a culture of spinal cord anterior horn (**Fig. 6**) and of cross-fractured growth cone (*gc* protoplasmic leaflet) and neurite *(n)* in a culture of superior cervical ganglion (**Fig. 7**). Note paucity of IMPs both on growth cones and in SPVs *(arrowheads)*. Concave faces of SPVs correspond to protoplasmic leaflet of plasmalemma. Rat tissues. *Circled arrow,* approximate shadowing direction for both figures. × 27,700 (**Fig. 6**) and × 63,500 (**Fig. 7**); *bars* 0.5 μm

In the mature neuron, by contrast, IMP densities are uniform throughout the axolemma (except for the presynaptic membrane). The phenomena observed in the growing neuron can be explained by the insertion of IMPs into the plasma membrane at the perikaryon and simultaneous rapid incorporation of particle-poor or particle-free plasmalemmal precursor (SPVs; Fig. 7) at the growth cone.

Conclusions

The various lines of evidence presented here consistently support the idea that the sprouting neuron exports, by rapid axoplasmic transport and at a high rate, a specific type of membrane from the perikaryon to the distal tip of the growing neurite. This plasmalemmal precursor is composed of newly synthesized phospholipid, certain types of glycoconjugates, and a very limited number of IMPs. The newly formed components are assembled in the perikaryon to form readily identifiable vesicles, here called SPVs. Most protein clusters (IMPs) of the axolemma appear to reach more distal sites of the neurite not by axoplasmic transport but by lateral diffusion within the axolemma.

Thus, a specific intracellular pathway has been identified for the plasmalemmal precursor to be inserted at the growth cone. This view is entirely consistent with a wide range of axoplasmic transport data obtained in different systems [18] as discussed extensively in this volume. The delayed appearance of IMPs in growing axolemma is the sign of a membrane maturation process in proximo-distal direction. It suggests, furthermore, that the plasmalemma itself has to be considered a potential pathway for the — most likely passive — transfer of membrane components from proximal to more distal cell regions, at least in the growing neurite where downhill gradients of such components exist.

Acknowledgments. The work presented in this chapter has profited from the invaluable assistance of Marian P. Johnson and Linda B. Friedman. Kathy Silberman has offered expert help with the completion of the manuscript. This research was supported by U.S. Public Health Service grant NS13466, NSF grant BNS 14071 and an I.T. Hirschl Career Scientist Award.

References

1. Bray D (1970) Surface movements during the growth of single explanted neurons. Proc Natl Acad Sci USA 65:905–910
2. Bray D (1973) Branching patterns of isolated sympathetic neurons. J Cell Biol 56:702–712
3. Hughes AF (1953) The growth of embryonic neurites. A study of cultures of chick neural tissue. J Anat 87:150–162
4. Lis H, Sharon N (1977) Lectins: their chemistry and application to immunology. In: Sela M (ed) The antigens, vol IV. Academic Press, London New York, pp 429–529
5. Maylié-Pfenninger M-F, Jamieson JD (1979) Distribution of cell surface saccharides on pancreatic cells. I. General method for preparation and purification of lectins and lectin-ferritin conjugates. J Cell Biol 80:69–76

6. Pfenninger KH (1979) Subplasmalemmal vesicle clusters: real or artifact? In: Rash JE, Hudson CS (eds) Freeze fracture: methods, artifacts and interpretations. Raven Press, New York, pp 71–80

7. Pfenninger KH (1980) Mechanism of membrane expansion in the growing neuron. Soc Neurosci Symp 6:661

8. Pfenninger KH, Bunge RP (1974) Freeze fracturing of nerve growth cones and young fibers. A study of developing plasma membrane. J Cell Biol 63:180–196

9. Pfenninger KH, Friedman LB (1982) Membrane biogenesis in the sprouting neuron: II. Autoradiographic analysis of the pathway to the cell surface of the plasmalemmal precursor (in preparation)

10. Pfenninger KH, Johnson MP (1982) Membrane biogenesis in the sprouting neuron: I. Selective transfer of newly-synthesized phospholipid into the growing neurite (in preparation)

11. Pfenninger KH, Maylié-Pfenninger M-F (1976) Differential lectin receptor content on the surface of nerve growth cones of different origin. Soc Neurosci Symp 2:224

12. Pfenninger KH, Maylié-Pfenninger M-F (1978) Characterization, distribution and appearance of surface carbohydrates on growing neurites. In: Karlin A, Vogel HJ, Tennyson VM (eds) Neuronal information transfer. Proc P & S Biomed Sci Symp. Academic Press, London New York, pp 373–386

13. Pfenninger KH, Malié-Pfenninger M-F (1981) Lectin labeling of sprouting neurons. I. Regional distribution of surface glycoconjugates. J Cell Biol 89:536–546

14. Pfenninger KH, Maylié-Pfenninger M-F (1981) Lectin labeling of sprouting neurons. II. Relative movement and appearance of glycoconjugates during plasmalemmal expansion. J Cell Biol 89:547–559

15. Salpeter MM, Bachmann L (1972) Autoradiography. In: Hayat MA (ed) Principles and techniques of electron microscopy. Biological applications, vol II. Van Nostrand Reinhold, New York, pp 220–278

16. Small R, Pfenninger KH (1980) Properties and maturation of axolemma in growing neurons. Soc Neurosci Symp 6:661

17. Small R, Pfenninger KH (1982) Components of the plasma membrane of growing axons. I. Spatial distribution of intramembranous particles (in preparation)

18. Weiss DG (ed) (1982) Axoplasmic transport. Springer, Berlin Heidelberg New York

19. Williams MA (1969) The assessment of electron microscopic autoradiographs. In: Barer R, Cosslett VE (eds) Advances in optical and electron microscopy, vol III. Academic Press, London New York, pp 219–272

The Role of Axonal Transport in the Growth of the Olfactory Nerve Axons

P. CANCALON, G.J. COLE, and J.S. ELAM [1]

Introduction

Numerous studies have analyzed changes in axonal transport that accompany nerve regeneration. These investigations have generally been carried out in vertebrate peripheral nerve or in lower vertebrate optic nerve, systems in which nerve injury is followed by rapid distal degeneration and regrowth of injured axons. (For review: see Grafstein and McQuarrie [7].) Findings to date have included changes in the amount and composition of rapidly transported molecules and changes in both rate of transport and composition of slowly transported molecules [6, 9, 14]. The significance of the observed changes in facilitating the processes of axonal outgrowth and functional reconnection is still poorly understood.

Recent studies in our laboratory have indicated that the garfish olfactory nerve may serve as a useful new model for the study of biochemical changes attendant to nerve regeneration. The nerve is characterized by unmyelinated and unbranched c-fibers (Fig. 1a) that provide a relatively homogeneous neuronal population as well as an unusually rich source of axonal plasma membrane [1]. The length (up to 30 cm) and lack of branching of the nerve provides a distance for monitoring axonal growth comparable to or greater than other peripheral nerves while the localization of terminals in the glomerular layer of the olfactory bulb facilitates investigation of subsequent synaptogenesis. In addition, the olfactory nerve has unusual regenerative capacity. Nerve section in a variety of species has been found to result in extensive death of the mature neurons followed by the appearance of a nearly normal number of new neuronal cells emanating from the germinal tissue of the olfactory mucosa [8]. In a limited sense the nerve can be considered to be a model for neuronal development in an adult animal. Even in the intact olfactory nerves, there is a continuing basal level of cellular turnover [8].

A combination of axonal transport labeling and morphological observations have been utilized to elucidate the pattern of olfactory nerve regeneration in the garfish. Results from experiments conducted at 21°C in which nerves were crushed 1.5 cm from the mucosal cell bodies and labeled with rapidly transported ^{35}S-methionine-labeled proteins yield a more complex pattern of regeneration (Fig. 2) [2]. A small

1   Department of Biological Science, Florida State University, Tallahassee, FL 32306, USA

Axoplasmic Transport in Physiology and Pathology
(ed. by D.G. Weiss and A. Gorio)
© Springer-Verlag Berlin Heidelberg 1982

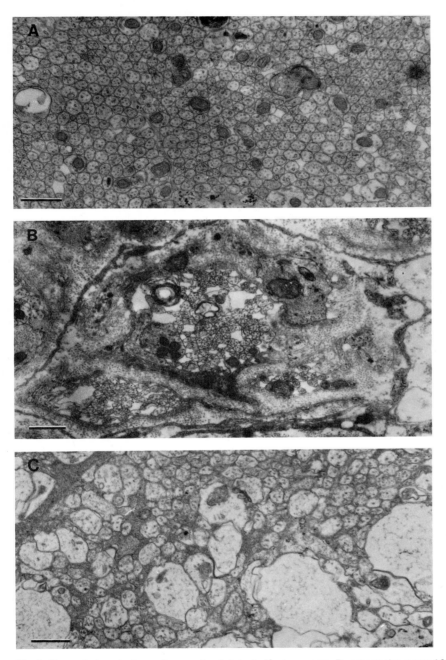

Fig. 1. Transmission electron micrographs. **A** Intact olfactory nerve, **B** regenerating nerve 102 d after crush (phases 1 and 2). Section taken 12 cm from the mucosa. **C** Regenerating nerve 189 d after crush (phase 3), section taken 5 cm from the mucosa. *Bar* 1 μm

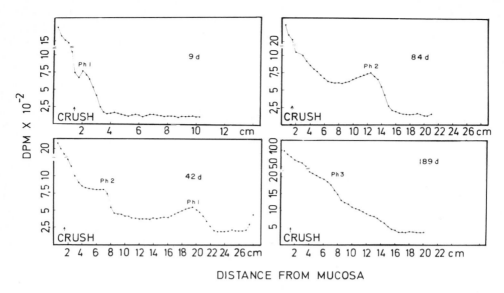

DISTANCE FROM MUCOSA

Fig. 2. Profiles of radioactivity determined by the accumulation of axonally transported ^{35}S-methionine labeled proteins in the rapidly growing fibers (phase 1), in the intermediate (phase 2), and slowly regenerating fibers (phase 3) at 9, 42, 84, and 189 days postoperatively. (From [3])

population of fibers (Fig. 1b) estimated by electron microscopy to be 3–5% of the original population grows through the crush site with short latency (6 days) and elongates at the unusually rapid rate of 5.8 mm/day (phase 1). The minimal latency suggests that these fibers arise from cells that have survived nerve crush and have regenerated their original axons. A second equally small population of fibers (Fig. 1b) emerges from the crush site at 13 days post crush and grows along the axon at a rate of 2.1 mm/day (phase 2). At a time corresponding to the arrival of the phase 1 fibers into the bulb area, the rate of elongation of this second neuronal population slows to 1.6 mm/day. This velocity decrease might be triggered by a feedback effect induced by the contact between the fastest growing fibers and their target. Since previous studies have suggested that a low rate of cell turnover occurs in uninjured olfactory nerve, we have hypothesized that the two initial populations of fibers may arise from olfactory cells that were immature or in the process of axonal outgrowth at the time of injury. The majority of the olfactory axons (50–70% of original) began to cross the crush site after 30 days and continued to appear for up to 200 days after crush (phase 3), reflecting the ongoing replenishment of the olfactory cell populations (Fig. 1c). The difference in the initial growth rates of the three axonal populations might be explained by a differential conditioning effect induced by the injury. McQuarrie and Grafstein [11] have shown that nerve regeneration following a crush is accelerated if the nerve is already recovering from a previous crush. In the present system the speeding effect of the conditioning lesion on axonal elongation might be expected to act more strongly on the neurons already extending axons at the time of injury (phase 1) than on those appearing later.

Our initial efforts to characterize the properties of axonal transport in the regenerating olfactory nerve have focused on rapid transport, with particular emphasis on (a) amounts of transported label, (b) rapid transport rate, (c) subcellular distribution of transported molecules, and (d) molecular characteristics of transported proteins and glycoconjugates.

Amount of Transported Label

Normal and regenerating nerve mucosae were labeled topically with 200 μCi of ^{35}S-methionine or ^{3}H-glucosamine [3]. Survival times were chosen to allow rapidly transported molecules to extend the entire length of regenerating nerve segments.

The first two phases of regenerating axons represent only a very small number of fibers (3–5% each), but, following labeling with labeled amino acids or carbohydrates each fiber contains two to three times more rapidly transported radioactivity than a fiber of the quantitatively larger third regenerating phase. Similar results have been obtained in the goldfish optic nerve where, following a conditioning lesion, regeneration after a testing injury proceeds at a faster rate and the amount of rapidly transported protein shows a 70% increase [11]. These results show a correlation between the amount of material transported and the rate of elongation, presumably reflecting the need for greater amounts of material to support more rapid growth.

Rates of Transport

Rates of fast axonal transport were estimated by determining transport peaks after different periods of transport in phase 1 or combined phase 1-phase 2 fibers. In agreement with numerous other studies [7], the most rapid rate of transport was found to be identical to that occurring in intact nerve [3]. As suggested above, nerve growth appears to be accomplished through increased amounts of protein in the transport phase rather than increased rates of delivery.

Preliminary studies of the rate of slow flow in the three phases of regenerating fibers indicate that the velocity of slow flow is identical in the different axonal populations. The rate of slow flow is approximately three times faster than the value measured in an intact nerve. The relationship between slow flow and nerve regeneration is presently being studied and it appears that the rate of slow flow must always be equal to or greater than the rate of nerve elongation. It can be hypothesized that in fully conditioned axons, where protein synthesis is maximum, the rate of slow flow becomes the elongation limiting factor. In less conditioned fibers other limiting factors such as the availability of newly synthesized molecules might be responsible for elongation rates inferior to the rates of slow flow.

Subcellular Distribution of Rapidly Transported Proteins

Electrophysiological and anatomical studies have pointed to the immaturity of the most rapidly regenerating olfactory fibers. Kiyohara and Tucker [10] have shown that during regeneration of the pigeon olfactory nerve three functionally distinct periods could be determined. During the first two weeks after the injury no electrophysiological activity can be measured following inhalation of an odorant by the animal. During the second two weeks a good response is obtained after the first, but not after subsequent inhalations, indicating that the nerve cannot respond to prolonged stimulation. After 30 days a normal response is obtained. Recently, using freeze fracture analysis of the axolemma of the regenerating bullfrog axons, Small and Pfenninger [15] have revealed the presence of an exponential proximo-distal decrease in the number of intramembranous particles which appear to be linked with the sodium channels. These studies suggest a possible delay in functional protein insertion that might be reflected in a lower buoyant density for the axonal membrane. In our studies we have found that the subcellular distribution of rapidly transported protein label in both phase 1 and 2 regenerating olfactory fibers shows a small but significant shift toward lower density membrane (Table 1). This subfraction has previously been shown to be enriched in plasma membrane markers [1] and the shift in label may reflect greater affiliation of transported protein with immature (low density) plasma membrane. Some shift in chemically assayed nerve protein to the lower density membrane fractions is also observed but is difficult to interpret in light of possible major glial contributions to the fraction. An increase in transported label in the "soluble" fraction is also observed, suggesting a possible role of cytoplasmic or loosely bound membrane proteins in the regenerative process.

Table 1. Subcellular distribution of radioactivity in intact and regenerating nerves (Cancalon and Elam [3])

	% of total radioactivity [a]		
	1st-phase terminals	1st-phase fibers and 2nd-phase terminals	Fast transport in intact nerve [b]
Soluble	16.5 ± 0.8 [c]	17.4 ± 1.7 [c]	11.3 ± 0.8
Particulate	82.6 ± 6.3	80.5 ± 4.8 [c]	89.6 ± 4.7
Total membrane	69.4 ± 2.3	68.7 ± 7.3	64.7 ± 3.0
Plasma membrane (20% sucrose)	25.5 ± 2.7 [c]	24.6 ± 3.1 [c]	14.5 ± 0.6
Plasma membrane (30% sucrose)	32.7 ± 1.3 [c]	31.4 ± 3.3 [c]	40.7 ± 1.9

[a] % of total TCA-insoluble radioactivits recovered in each area of the nerve considered (values are means ± S.D. of six experiments)

[b] Area of the intact nerve containing the peak of rapidly moving radioactivity in an unoperated animal [1]

[c] Significantly different ($P < 0.05$) from fast transport values in intact nerve

Composition of Transported Molecules

Transported proteins of phase 1 and phase 2 regenerating fibers were labeled with ^{35}S-methionine and compared with those of intact nerve following undimensional PAGE. A generally similar pattern of labeled protein was observed (Fig. 3). However resolution in this gel system would not reveal changes in a limited number of individual proteins, as has been observed in several other regenerating nerves [6, 9, 14]. In view of the possible special function of transported glycoconjugates in cell recognition processes during nerve regeneration, we have also examined labeling of the oligosaccharide chains associated with the rapidly transported molecules [4]. Transported molecules in phase 3 fibers were labeled with ^{3}H-glucosamine, then subjected to pronase digestion and subsequent procedures for isolation of glycopeptide and glycosaminoglycan (GAG) chains. Results show changes in the composition of transported glycopeptides. One is a shift to greater transport labeling of the lower molecular weight (dialyzable) sized molecules (Table 2). This type of oligosaccharide chain has

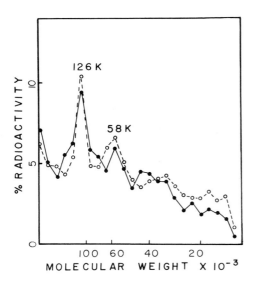

Fig. 3. Distribution of rapidly transported ^{35}S-methionine labeled proteins in 10% SDS polyacrylamide gels of the area of a regenerating nerve containing the first phase of regenerating fibers (- ● -) and of the area of an intact nerve containing the molecules transported at the rapid rate (- ○ -). (From [3])

Table 2. Relative distribution of ^{3}H-glucosamine in axonally transported glycopeptides and glycosaminoglycans (GAGs)

	% Recovered radioactivty		
	Glycopeptides		GAGs
	Non-dialyzable	Dialyzable	
Intact	65.5 ± 3.0	23.9 ± 3.0	10.6 ± 1.5
Regenerating	46.3 ± 5.2	46.0 ± 5.2	7.7 ± 1.3

Table 3. Con A-binding of transported glycopeptides in control
and regenerating nerve

	% of total dialyzable or non-dialyzable radioactivity binding Con A (eluted with methylglucoside)	
	Non-dialyzable	Dialyzable
Intact nerve	12.2 ± 1.8	17.5 ± 3.5
Regenerating nerve	23.4 ± 3.3 [a]	32.6 ± 6.7 [b]

[a] Significantly different from intact nerve at $p < 0.01$
[b] Significantly different from intact nerve at $p < 0.025$

been correlated with cell growth in non-neural systems [2] and may reflect a role in altered cell surface interactions or growth regulation per se. Studies on developing neurons of rat cerebellum [16] and brain [5] have indicated transient increases in cell surface glycoconjugates that bind Concanavalin A (Con A). As a test of whether similar increases may occur in regenerating nerve, transported labeled glycopeptides were chromatographed on columns of Concanavalin A-Sepharose. As can be seen in Table 3, regeneration is characterized by significant increases in the proportion of both dialyzable and larger non-dialyzable glycopeptides binding Con A. When the increased proportion of the dialyzable molecules is taken into account, an overall three to fivefold increase in the proportion of low molecular weight-Con A binding molecules can be estimated. Preliminary compositional analysis indicates the presence of polymannosyl type glycopeptides typically showing high Con A affinity.

Although glycosaminoglycans also have been shown to undergo compositional changes in association with cell growth [13], the present studies show no significant difference in the proportion of transported label associated with GAGs in regeneration (Table 2). However, preliminary analysis suggests some change in the proportion of individual types of GAG.

Conclusion

While several model systems are available for the study of axonal transport in regenerating nerve, the garfish olfactory nerve appears to hold promise of providing additional insight into the problem. In studies already completed, the length and homogeneity of the nerve has permitted elucidation of both subcellular shifts in transported molecules and compositional changes potentially relevant to cell surface dynamics. The preparation also allows precise analysis of rates and composition of slow flow in regenerative growth and analysis of all phases of transport during the period of reconnection at the olfactory bulb. Finally, the olfactory nerve preparation provides the unique opportunity to study growth of neurons from germinal tissue in an adult animals.

Acknowledgments. This study was supported in part by the National Spinal Injury Foundation and by NIH grants NS 17198 and NS 11456.

References

1. Cancalon P, Beidler LM (1975) Distribution along the axon and into various subcellular fractions of molecules labeled with ^{3}H-leucine and rapidly transported in the garfish olfactory nerve. Brain Res 89:225–244
2. Cancalon P, Elam JS (1980) Study of regeneration in the garfish olfactory nerve. J Cell Biol 84:779–794
3. Cancalon P, Elam JS (1980) Rate of movement and composition of rapidly transported proteins in regenerating olfactory nerve. J Neurochem 35:889–897
4. Cole GJ, Elam JS (1981) Axonal transport of glycoproteins in regenerating olfactory nerve: Enhanced glycopeptide Concanavalin A-binding. Brain Res 222:437–441
5. DeSilva NS, Gurd JW, Schwartz C (1979) Developmental alteration of rat brain synaptic membrane. Reaction of glycoproteins with plant lectins. Brain Res 165:283–293
6. Giulian D, Des Ruisseaux H, Cowburn D (1980) Biosynthesis and intraaxonal transport of proteins during neuronal regeneration. J Biol Chem 255:6469–6501
7. Grafstein B, McQuarrie IG (1978) Role of nerve cell body in axonal regeneration. In: Cotman CW (ed) Neuronal plasticity. Raven Press, New York, pp 155–195
8. Graziadei PPC, Moti-Graziadei GA (1978) Continuous nerve cell renewal in the olfactory system. In: Jacobson M (ed) Handbook of sensory physiology, vol IX: Development of sensory systems. Springer, Berlin Heidelberg New York, pp 55–82
9. Hoffman PN, Lasek RJ (1980) Axonal transport of the cytoskeleton in regenerating motoneurons: constancy and changes. Brain Res 202:327–333
10. Kiyohara S, Tucker D (1978) Activity of new receptors after transection of the primary olfactory nerve in pigeons. Physiol Behav 21:987–994
11. McQuarrie IG, Grafstein B (1978) Protein synthesis and fast transport in regenerating goldfish retinal ganglion cells: effect of a conditioning lesion. Soc Neurosci Abstr 4:533
12. Muramatsu T, Koide N, Ceccarini C, Atkinson PH (1976) Characterization of mannose-labeled glycopeptides from human diploid cells and their growth dependent alterations. J Biol Chem 251:4673–4679
13. Ninomiya Y, Hata R-I, Nagai Y (1980) Glycosaminoglycan synthesis by liver parenchymal cell clones in culture and its change with transformation. Biochim Biophys Acta 629:349–358
14. Skene JHP, Willard M (1981) Axonally transported proteins associated with axon growth in rabbit central and peripheral nervous system. J Cell Biol 89:96–103
15. Small KR, Pfenninger KH (1980) Properties and maturation of axolemma in growing neurons. Soc Neurosci Abstr 6:661
16. Zanetta JT, Roussel G, Ghandour MS, Vincendon G, Gombos G (1978) Postnatal development of rat cerebellum: massive and transient accumulation of Concanavalin A binding glycoproteins in parallel fiber axolemma. Brain Res 142:301–319

Changes in Fast-Transported Protein in Regenerating Axons: Essential or Incidental?

MARK A. BISBY [1]

Introduction

Following axotomy and during subsequent axon regeneration, the cell bodies of peripheral neurons undergo structural and metabolic changes known as "chromatolysis", including an alteration in protein synthesis patterns of the cell bodies. Synthesis of transmitter-related enzymes decreases while that of structural proteins increases [15]. Changes in the composition of fast transported protein might be expected when the axon is elongating instead of transmitting to the target cell; thus, changes in the transport of specific transmitter-related enzymes have been reported [17, 21, 23, 24, 32]. The study of changes in the transport of total transported proteins might provide information about the functions of some of the transported proteins, and conversely, about the role of fast axonal transport in axon regeneration.

Motoneuron Axons

Rat sciatic nerves were crushed distally, and at various intervals L-[^{35}S]-methionine was injected into the lumbosacral spinal cord. A collection crush was placed around the sciatic nerve 10 mm proximal to the original crush. After allowing 6 h for synthesis of protein and transport into the nerve, the 3 mm nerve segment proximal to the collection crush was removed and processed for SDS-polyacrylamide gel electrophoresis and subsequent fluorography of the gel in order to detect the ^{35}S-labelled polypeptides.

The motoneuron axons regenerated promptly after a crush injury: within about 2 days the growing axons penetrated through the injured region and the fastest growing axons regenerated at 4 mm/day^{-1}. Re-innervation began 21–28 days after injury, and by 42 days movements in the leg supplied by the injured nerve were grossly normal and all muscles were innervated.

1 Dept. of Medical Physiology, Faculty of Medicine, University of Calgary, Calgary, Alta, T2N 4N1, Canada

Axoplasmic Transport in Physiology and Pathology
(ed. by D.G. Weiss and A. Gorio)
© Springer-Verlag Berlin Heidelberg 1982

During regeneration, changes occurred in the labelling of certain fast-transported polypeptides. In common with other studies on axonal transport during regeneration [25, 30] no new polypeptides appeared during regeneration, nor did "normal" polypeptides disappear. Overall, there were surprisingly few changes in the composition of labelled transported protein, but the changes did parallel the events of regeneration; significant changes occurred by 3 days and by 42 days the composition of labelled protein had returned to normal (Fig. 1) [5].

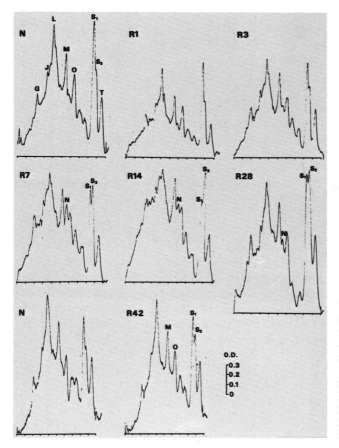

Fig. 1. Changes in composition of axonally-transported polypeptides during regeneration of rat sciatic nerve motoneurons. Densitometer scan of a representative fluorograph of a slab SDS-PAGE gel containing extract from normal nerve *(N)* and nerves various days after injury *(R 1-45)*. [5]

The major and most consistent change was an increase in labelling of a band of MW 23,000 daltons which we designated "S_2", perhaps the same as "GAP 23" [1]. The change was most evident when compared to a neighbouring band "S_1" (Fig. 2). In subsequent studies the S_2/S_1 ratio was used as an index of changes in fast transport.

Many chromatolytic changes are reversed when the regenerating motoneuron axons succeed in re-innervating the muscle, but if regeneration does not occur the changes are not reversed [10, 19]. When the sciatic nerve was resected, no re-innervation of

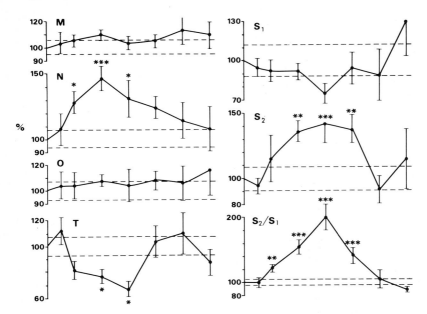

Fig. 2. Time-course of changes. Normalized curves for the relative density of the indicated labelled bands during regeneration: some bands show significant changes (*N, T, S₂*), others do not. *Horizontal axis,* number of days elapsed since nerve injury; *vertical axis,* 100% is the value of relative band density in normal nerves. [5]

distal limb musculature occurred. Under these circumstances the S_2/S_1 ratio remained elevated up to at least 100 days after nerve injury. Thus the S_2/S_1 ratio seems to be sensitive to axonal regeneration or re-innervation of the target cells. It might be under control of signals emanating from the target tissue and retrograde-transported to the cell body when axons re-establish innervation.

Sensory Neurons

Since transported proteins in motor and sensory axons are similar [1a, 3, 28] I wondered whether the change in the S_2/S_1 ratio was a phenomenon unique to regenerating motoneurons. Fourteen days after the sciatic nerve was crushed ^{35}S-methionine was injected into the L5 dorsal root ganglion. An increase in S_2/S_1 ratio occurred in both central and peripheral processes of the sensory axon (Fig. 3) [6]. There was no evidence for specific "routing" into only the regenerating branch. While there is a quantitative difference in transport into central and peripheral axon processes [22] there is no evidence for selective transport into only a single process.

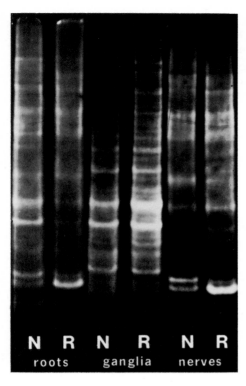

Fig. 3. Sensory neurons. Representative fluorograph from normal *(N)* and regenerating *(R)* dorsal roots, L5 dorsal root ganglion and sciatic nerves 14 days after sciatic nerve injury. Bands S_1 and S_2 are the two prominent adjacent bands at the lower end of the fluorograph. Note the increased labelling of band S_2 in both regenerating roots and nerves

Significance of Observed Changes

The similarity of the transported polypeptides, and the changes that occur during regeneration, in neurons with different neurotransmitter systems and physiological functions, implies a common role for transported proteins. Contrary to what might expected, there is an initial decrease in quantity of labelled protein and glycoprotein transported after injury, and a recovery to normal levels during regeneration [4, 8, 11]. Studies on the transport of acetylcholinesterase show a similar pattern [12, 17, 21, 24]. The transported protein in regenerating axons is preferentially incorporated into the growing axons [4, 13] and transported glycoprotein is inserted into axonal membrane mainly at the growth cone but also along the length of new axon [16, 29].

The work of Hoffman and Lasek [18] reveals the importance of *slow* axonal transport in providing the structural matrix for axon outgrowth. *Fast* axonal transport probably supplies relatively small quantities of membrane-bound protein to the newly-formed axons, these required proteins being similar in all neurons, and involved in common functions such as impulse conduction and ion pumping. Since rapidly transported proteins turn over rapidly, as shown by the time-course of their subsequent retrograde transport [2, 7], and autoradiographic studies on the axon terminals [9], the neuron renews its fast-transported proteins almost daily, so that a change to the regenerating state provokes only slight changes in the composition of transported protein. Axotomy reduces the membrane area of the axon and might lead to reduced demand for fast-transported proteins. As the axon regenerates, the demand would increase towards normal levels. Of course, this argument supposes that the cell body "knows" how long its axon is.

What is the significance of the changes that *do* occur during regeneration? Perhaps they regulate regeneration. The "conditioned lesion" phenomenon supports this idea. If axons are injured (conditioning lesion), allowed to regenerate and then injured again more proximally, regeneration from the second (test) injury occurs more promptly and rapidly [20]. It is supposed that the conditioning lesion changes the neuron to a regenerating status so that when the test injury is made the neuron is ready to regenerate without delay. Studies on the time-course of the conditioning lesion effect [14] show that it will occur with a conditioning-test interval of as short as 2 days, much too early for changes in slow transport ($1-4$ mm/day^{-1}) to reach the lesion site, but within the period when changes in fast-transported protein become significant (Fig. 1).

Central Nervous System (CNS) Axons

CNS neurons normally do not regenerate: do they respond to axotomy differently from peripheral neurons? In rat optic nerves we found no increase in S_2/S_1 ratio after lesioning. In fact, within 4 days the quantity of transported protein in the optic nerve decreased considerably, and drastic changes in the composition of labelled proteins occurred. These were not, however, due to alterations in axonally transported protein, since the new labelled polypeptides present in the lesioned nerve were the same as those seen in normal nerve when transport was blocked by colchicine [26]. We concluded that they represented glial protein synthesized from diffused label. The rapid decline in transport following optic nerve injury might be due to anoxia of the proximal nerve stump caused by interruption of its blood supply. At present, therefore, the absence of an increase in S_2/S_1 ratio in rat optic nerve after injury cannot be correlated with its failure to regenerate. However, Skene and Willard reported that changes in fast-transported protein associated with regeneration are absent from injured rabbit optic nerve axons [27], consistent with the correlation between regenerative capacity of axons and changes in axonal transport.

Conclusion

There is at present no firm evidence for or against the hypothesis that changes in fast transport regulate regeneration. Such regulation probably would not extend to the actual initiation of axonal sprouting and the formation of the growth cone.

In cultured neurons initiation of growth cone activity after axon injury occurs very rapidly (3–10 min [31]), probably too soon for an injury signal to reach the cell body, to initiate changes in protein synthesis, and for the alteration in protein synthesis to reach the axon terminals. Alternatively, changes in fast axonal transport may not be regulatory but only supportive of regeneration, providing the components required for construction of functional axonal membrane behind the advancing growth-cone.

Acknowledgments. Supported by grants from the Medical Research Council of Canada and the Alberta Heritage Foundation for Medical Research.

References

1. Baitinger C, Levine J, Simon C, Skene P, Willard M (1982) Characteristics of axonally transported proteins. In: Weiss DG (ed) Axoplasmic transport. Springer, Berlin Heidelberg New York, pp 110–120
1a. Barker JL, Neale JH, Gainer H (1976) Rapidly transported proteins in sensory, motor and sympathetic nerves of the isolated frog nervous system. Brain Res 105:497–515
2. Bisby MA (1976) Orthograde and retrograde axonal transport of labelled protein in motoneurons. Exp Neurol 50:628–640
3. Bisby MA (1977) Similar polypeptide composition of fast-transported proteins in motor and sensory axons. J Neurobiol 8:303–314
4. Bisby MA (1978) Fast axonal transport of labelled protein in sensory axons during regeneration. Exp Neurol 61:281–300
5. Bisby MA (1980) Changes in the composition of labelled protein transported in motor axons during their regeneration. J Neurobiol 11:435–445
6. Bisby MA (1981) Axonal transport in the central axon of sensory neurons during regeneration of their peripheral axon. Neurosci Lett 21:7–11
7. Bisby MA, Bulger VT (1977) Reversal of axonal transport at a nerve crush. J Neurochem 29:313–320
8. Bulger VT, Bisby MA (1978) Reversal of axonal transport in regenerating nerves. J Neurochem 31:1411–1418
9. Droz B (1975) Synthetic machinery and axoplasmic transport: maintenance of neuronal connectivity. In: Tower DB (ed) The nervous system, vol I. Raven Press, New York, pp 111–127
10. Engh CA, Schofield BH, Doty SB, Robinson RA (1971) Perikaryal synthetic function following reversible and irreversible axon injuries as shown by radioautography. J Comp Neurol 142:465–480
11. Frizell M, Sjöstrand J (1974) The axonal transport of [^{3}H]-fucose labelled glycoproteins in normal and regenerating peripheral nerves. Brain Res 78:109–123
12. Frizell M, Sjöstrand J (1974) Transport of proteins, glycoproteins and cholinergic enzymes in regenerating hypoglossal nerves. J Neurochem 22:845–880
13. Forman D, Berenberg RA (1979) Regeneration of motor axons in the rat sciatic nerve studied by labelling with axonally transported radioactive proteins. Brain Res 156:213–225
14. Forman D, McQuarrie IG, Labore FW, Wood DK, Stone LW, Braddock CH, Fuchs DA (1979) Time-course of the conditioning lesion effect on axonal regeneration. Brain Res 182:180–185
15. Grafstein B, McQuarrie IG (1978) Role of nerve cell body in axonal regeneration. In: Cotman CW (ed) Neuronal plasticity. Raven Press, New York, pp 155–195

16. Griffin JW, Price DL, Drachman DB, Morris J (1981) Incorporation of axonally transported glycoproteins into axolemma during nerve regeneration. J Cell Biol 88:205–214
17. Heiwall PO, Dahlström A, Larsson PA, Bööj S (1979) The intraaxonal transport of acetylcholine and cholinergic enzymes in rat sciatic nerve during regeneration after various types of axonal trauma. J Neurobiol 10:119–136
18. Hoffman PN, Lasek RJ (1980) Axonal transport of the cytoskeleton in regenerating motor neurons: constancy and change. Brain Res 202:317–334
19. Lieberman AR (1974) Some factors affecting retrograde neuronal responses to axonal lesions. In: Bellairs R, Gray EG (ed) Essays on the nervous system. Clarendon Press, Oxford, pp 71–105
20. McQuarrie IG, Grafstein B, Gershon MD (1977) Axonal regeneration in the rat sciatic nerve; effect of a conditioning lesion and of dbCAMP. Brain Res 132:443–453
21. O'Brien RAD (1978) Axonal transport of acetylcholine, choline acetyltransferase and cholinesterase in regenerating peripheral nerve. J Physiol (London) 282:91–103
22. Ochs S, Erdman J, Jersild RA, McAdoo V (1978) Routing of transported materials in the dorsal root and nerve fiber branches of the dorsal root ganglion. J Neurobiol 9:465–481
23. Oshiro S, Fujiwara M, Osumi Y (1978) Axonal transport of norepinephrine and choline acetyltransferase in regenerating sciatic nerve of rat. Exp Neurol 62:159–172
24. Schmidt RE, McDougal DB Jr (1978) Axonal transport of selected particle-specific enzymes in rat sciatic nerve in vivo and its response to injury. Neurochem 30:527–536
25. Perry GW, Wilson SL (1980) Protein synthesis and axonal transport following peripheral nerve damage. Soc Neurosci Abstr 6:94
26. Redshaw JD, Bisby MA (1981) Fast axonal transport of protein in retinal ganglion cell axons following axotomy. Fed Proc 40:302
27. Skene JHP, Willard M (1980) Two axonally transported proteins associated with growing axons in rabbits. Soc Neurosci Abstr 6:687
28. Stone DL, Wilson GC (1979) Quantitative analysis of proteins rapidly transported in ventral horn motoneurons and bidirectionally from dorsal root ganglia. J Neurobiol 10:1–12
29. Tessler A, Autilio-Gambetti L, Gambetti P (1980) Axonal growth during regeneration – quantitative autoradiographic study. J Cell Biol 87:197–203
30. Theiler RF, McClure WO (1978) Rapid axoplasmic transport of proteins in regenerating sensory nerve fibers. J Neurochem 31:433–447
31. Wessels NK, Johnson SR, Nuttall RP (1978) Axon initiation and growth cone regeneration in cultured motor neurons. Exp Cell Res 117:335–345
32. Wooten GF, Coyle JT (1973) Axonal transport of catecholamine synthesizing and metabolizing enzymes. J Neurochem 20:1361–1371

Recovery of Axonal Transport of Acetylcholinesterase in Regenerating Sciatic Nerve Precedes Muscle Reinnervation

LUIGI DI GIAMBERARDINO[1,3] , JEAN-YVES COURAUD[1] ,
RAYMONDE HÄSSIG[1] , and ALFREDO GORIO[2]

Recent reports have shown that the effect of nerve trauma on the axonal transport of acetylcholinesterase (AChE) may differ depending on the type of injury. While the axonal transport of AChE was seen to decrease drastically during the 2—3 days following nerve injury, whatever the injury [3, 5, 7], the recovery which followed this initial drop was seen to take place only after nerve crush [5, 7] or nerve freezing [5] but not after nerve section and very little after nerve ligature [5].

These observations raise the question of which signal induces the recovery of AChE transport in a crushed nerve and why this signal is missing after nerve section. A tentative answer to this question was proposed by O'Brien [7], who suggested that the reinnervation of the denervated muscles might be the event which triggers the recovery of AChE transport. This view is not supported by the work presented here. In this work the time course of muscle reinnervation, monitored by measuring its AChE concentration, was correlated with the recovery of AChE axonal transport after crushing the sciatic nerve of rats. The results obtained indicate that axonal transport of AChE is restored before muscle reinnervation.

Our experiments were carried out on 180—200 g albino Sprague-Dawley rats. The sciatic nerve was crushed just before the gluteal branching. At several time intervals after nerve crush, the EDL (M. extensor digitorum longus) concentration of AChE and the axonal transport of this enzyme were studied (see legends to Figs. 1 and 2 for experimental details). As a measure of AChE axonal transport, we used its accumulation over a 24 h period at a test section (Fig. 2). This parameter is probably the balance between the anterograde and the reverse flow-rate of this enzyme at the accumulation site (see [2]) and it can be taken as a good indicator of the variations of the anterograde axonal transport.

The time course of the variation of EDL AChE content (Fig. 1), shows that the AChE concentration declines at first very rapidly, up to the 5th day after nerve crush, then more slowly, between day 5 and 15. An identical pattern was reported by others [8]. After day 15 the concentration starts recovering and reaches its maximum value at day 30. The control value was not reached, however, even at 60 days.

1   Déparment de Biologie, Commissariat à l'Energie Atomique, 91191 Gif-sur-Yvette, France
2   FIDIA Research Laboratories, Abano Terme, Italy
3   INSERUM U. 153, 17 rue du fer à moulin, 75005 Paris, France

Axoplasmic Transport in Physiology and Pathology
(ed. by D.G. Weiss and A. Gorio)
© Springer-Verlag Berlin Heidelberg 1982

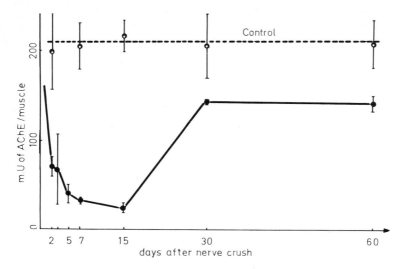

Fig. 1. AChE content in the rat EDL after a sciatic nerve crush. EDL muscles were taken at several time intervals after the sciatic nerve crush. Whole muscles were ground in liquid nitrogen. An aliquot of the resulting powder was further homogenized in 10% (w/v) of extraction buffer (see legend to Fig. 2) in the presence of 40 µg/ml leupeptin and 20 µg/ml pepstatin as additional protease inhibitors. The assay of AChE activity was performed as indicated in the legend to Fig. 2. Control muscles were obtained from nonoperated, same aged animals. Each value is the mean ± S.E.M. of three determinations

The kinetics of EDL AChE recovery (Fig. 1) follow very closely that of the reinnervation of this muscle, as monitored by the appearance of complex end plate potentials (epps [4, 5]; see also Gorio, this volume). Hence, the recovery of AChE concentration can be taken as a reliable reinnervation index in our experimental conditions.

The time course of the variation of anterograde transport of AChE after nerve crush is shown in Fig. 2. The accumulation rate of AChE at the test section starts decreasing 24 h after nerve crush, reaches its minimum value at day 4, and rises thereafter. The control value, measured in uncrushed nerves of coeval rats, is attained between day 30 and 60.

From these results it appears that recovery of AChE transport occurs at least 10 days before the recovery of AChE concentration in the EDL. It thus seems rather unlikely that reinnervation itself could be the signal which prompts the resumption of AChE transport. Hence the event which triggers the recovery of axonal transport must precede reinnervation and might well be an important event allowing successful nerve regeneration. Such an event is apparently missing in a sectioned nerve which is definitely less able than a crushed nerve to regenerate and achieve satisfactory reinnervation. As to the nature of this event we can only guess; a good candidate could be the encounter of an axonal sprout, budding from the injured region, with an old basal lamina channel [9] once filled with a now degenerated axon.

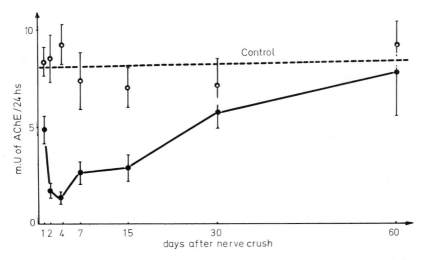

Fig. 2. Evolution of the anterograde accumulation rate of AChE in crushed rat sciatic nerve. At several time intervals after nerve crush the accumulation rate of AChE was measured in the proximal stump by the nerve-section method [1]. To that purpose the nerve was sectioned 3 to 5 mm proximal to the nerve crush and the enzyme was left to accumulate at the section. The AChE activity was assayed 24 h later in the 2 mm nerve segment proximal to the section. Each 2 mm nerve segment was homogenized in 200 μl of extraction buffer (1 M NaCl, 0.01 M phosphate buffer pH 7.0,5 mM EGTA, 1% Triton X-100, 1 mg/ml bacitracine, 2 mM benzamidine). After centrifugation (20,000 × g, 20 min), AChE activity was assayed in an aliquot of the supernatant using 0.75 mM acetylthiocholine as substrate and 0.1 mM isoOMPA as pseudocholinesterase inhbitor. The anterograde accumulation rate of AChE was defined as the quantity of enzyme, beyond that normally present in the nerve, accumulated over a 24 h period at the 2 mm nerve segment proximal to the section. Control accumulation rates were measured in sciatic nerves of uncrushed rates of the same age. Each *point* represents the mean ± S.E.M. of at least three animals

References

1. Couraud JY, Di Giamberardino L (1980) Axonal transport of the molecular forms of acetylcholinesterase in chick sciatic nerve. J Neurochem 35:1053–1066
2. Couraud J-Y, Di Giamberardino L (1982) Axonal transport of the molecular forms of acetylcholinesterase. Its reversal at a nerve transection. In: Weiss DG (ed) Axoplasmic transport. Springer, Berlin Heidelberg New York, pp 144–152
3. Frizell M, Sjöstrand J (1974) Transport of proteins, glycoproteins and cholinergic enzymes in regenerating hypoglossal neurons. J Neurochem 22:845–850
4. Gorio A, Carmignoto G, Facci L, Finesso M (1980) Motor nerve sprouting induced by ganglioside treatment. Possible implications for gangliosides on neuronal growth. Brain Res 197: 236–241
5. Heiwall PO, Dahlström A, Larsson PA, Bööj S (1979) The intraaxonal transport of acetylcholine and cholinergic enzymes in rat sciatic nerve during regeneration after various types of axonal trauma. J Neurobiol 10:119–136
6. McArdle JJ (1975) Complex end plate potentials at the regenerating neuromuscular junction of the rat. Exp Neurol 49:629–638

7. O'Brien RAD (1978) Axonal transport of acetylcholine, choline acetyltransferase and cholinesterase in regenerating peripheral nerve. J Physiol (London) 282:91–103
8. Ranish NA, Kiauta T, Dettbarn WD (1979) Axotomy induced changes in cholinergic enzymes
   in rat nerve and muscles. J Neurochem 32:1157–1164
9. Thomas PK (1964) Changes in the endoneural sheaths of peripheral myelinated nerve fibres
   during Wallerian degeneration. J Anat 98:175–182

The Role of Axoplasmic Transport in the Restoration of Synaptic Transmission and in the Process of Sprouting During Nerve Regeneration

ALFREDO GORIO [1]

Central and Peripheral Reaction

After lesion of a peripheral nerve there is a complex of histometabolic reactions that occurs in the cell body and dendritic tree. Some changes are even observable at the light microscopic level and were termed chromatolysis. In summary, the reaction is characterized by dissolution of the basophilic granules, rounding of the perikaryon, enlargement of the nucleolus, and dispersal of the Nissl substance. The latter corresponds to the disaggregation of cytoplasmic ribosomal clusters and degranulation of the rough endoplasmic reticulum. This process is inhibited if actinomycin D is injected at the time of the axonal damage or within 9 h, suggesting the involvement of RNA synthesis as a trigger for chromatolytic reaction [15]. It is a fact regenerating neurons produce more proteins, even at faster rates, as shown by many laboratories (see the preceding chapters in this volume).

The metabolic reactions in the soma are also accompanied by several modifications of the neuropil; for example, there is a loss of presynaptic nerve endings from the motoneuron soma and dendrites, which are somewhat recovered when the regeneration process is terminated [2, 15].

Nerve degeneration also has profound effects on the periphery which is deprived of its own innervation. The lack of neuronal trophism and mechanical activity affects several biophysical and metabolic parameters of the muscle. For instance, acetylcholine receptors are strictly located at the neuromuscular junction, but after denervation the sensitivity to acetylcholine (Ach) is spread over the entire muscle surface. This phenomenon, which has been named supersensitivity [1, 19], is due to the appearance of a new family of receptors at extrasynaptic sites [5, 6] The increase in Ach sensitivity is accompanied by both a drop in membrane resting potential, which can be as much as 20 mV, and muscle fibre atrophy. Within a week from denervation the muscle becomes autoexcitable, because spontaneous action potentials originate from the end plate region and propagate all over the surface, inducing muscle fibrillation. In addition, the pharmacological properties of the Na-channel are also changed; in fact, the action potentials are insensitive to tetrodotoxin [6].

---

1   Department of Cytopharmacology, Fidia Research Laboratories, Via Ponte della Fabbrica 3/A, 35031 Abano Terme (PD), Italy

Axoplasmic Transport in Physiology and Pathology
(ed. by D.G. Weiss and A. Gorio)
© Springer-Verlag Berlin Heidelberg 1982

Reinnervation Process and Axonal Transport

Many investigators have used the neuromuscular model to study the process of regeneration of the nervous system in an attempt to determine its specificity. The relatively simple organization of the neuromuscular junction and the possibility of correlating structure and function have made this preparation the ideal choice. Therefore, the literature available about this matter is extremely abundant and it widely covers the subject for mammalian and non-mammalian animals [10, 18, 21, 22].

From recent work by U.J. MacMahan et al. it appears that the growing tips of regenerating motoneurons find a specific recognition site in some components of the basal lamina present exclusively at the neuromuscular junction [20]. Therefore, it seems that growth of regenerating axons is blocked at specific sites and there the growing tip is converted into a releasing nerve terminal.

The model we have chosen for studying the influence of axonal transport in the synaptic restoration is the sciatic nerve — extensor digitorum longus (EDL) muscle preparation of the rat. Denervation was accomplished by crushing the nerve in the upper thigh and the reinnervation process monitored by electrophysiological and morphological means [8, 9]. Muscles were processed for either light or electron microscopic examination after the electrophysiological assessment.

The first contact between nerve and muscle occurs after 2 weeks and is monitored electrophysiologically by the recording of subthreshold end plate potentials [3] which become overthreshold within 24 h. Consequently, the mechanical activity of the muscle is quickly regained.

The early stages of reinnervation are characterized by thin unmyelinated axons leaving the intramuscular trunks of the nerve and reforming a functional neuromuscular junction at precisely the old sites. The nerve endings are very small, contain few synaptic vesicles, and resemble a transient state between growth cone and nerve terminal releasing neurotransmitter.

The process of synapse formation can be divided into several steps but basically it is composed of two phases: the first related to the formation of the neuromuscular contact and the second to the suppression of the redundant innervation. In a normally innervated muscle there is a one to one relationship between axons leaving the nerve branches and nearby end plates; this picture is altered however during reinnervation. In fact, before making synapses with the muscle fibres, growing motoneurons sprout collaterally to sequentially innervate several end plates. This sequence of events is characteristic of all the regenerating axons, and, therefore, an abundancy of contacts is made and a single neuromuscular junction is formed by several nerve endings placed over the same postsynaptic area. The number of polyinnervated synapses increases with time and reaches the maximum level 10 days after reinnervation. Then repression starts and within 2 months the redundant nerve terminals are repressed, leading to monoinnervated muscle fibres as in normal muscles (Figs. 1, 2, 3, 4).

The two phases are probably controlled by different types of stimuli; in the first phase growth, sprouting, and synapse formation are stimulated, while in the second one growth ceases and the elimination of superfluous nerve endings occurs, suggesting that signals must act in opposite ways with respect to the previous ones. In both cases

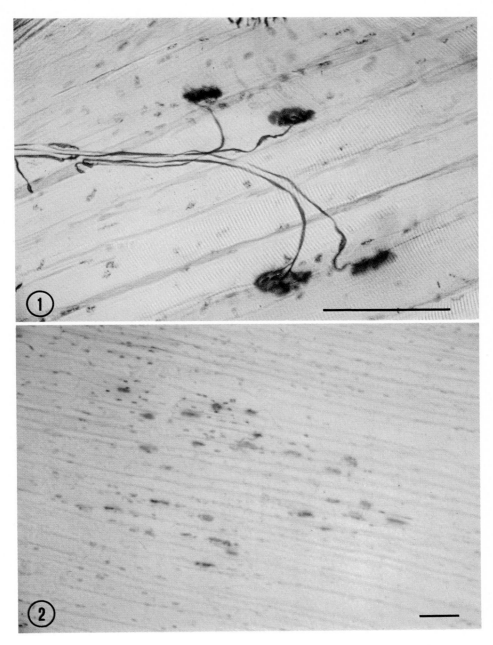

Fig. 1. Normal pattern of extensor digitorum longus muscle (EDL) innervation. 20 μm frozen sections of the muscle are stained with bromohindoxyl for the end plate acetylcholinesterase and silver impregnated for the axons. The same procedure was used for Figs. 2, 3 and 4. *Bar* 100 μm

Fig. 2. Section of a denervated muscle, which shows the total lack of silver impregnated axons. *Bar* 100 μm

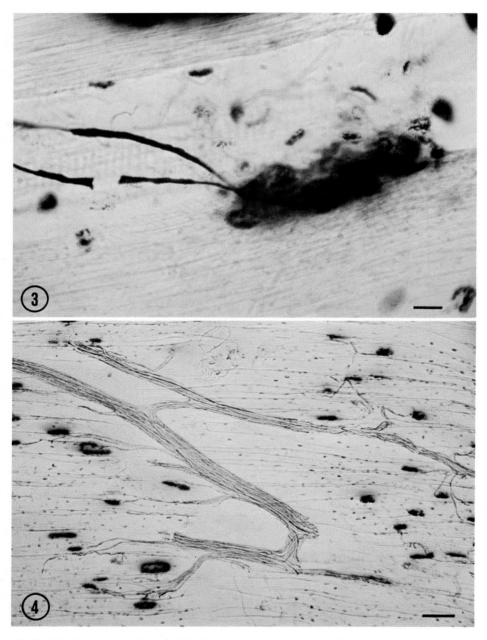

Fig. 3. A double innervated muscle fibre from a muscle 21 days after denervation. At this stage of reinnervation process about 60% of the muscle fibres are polyinnervated. *Bar* 10 μm

Fig. 4. Photomicrograph of a muscle 90 days after nerve crush, which shows the regained normal pattern of innervation. *Bar* 100 μm

the signals must be retrogradely transported to the cell body, indicating what is happening in the periphery so as to regulate the neuronal metabolic machinery. These signals are probably regulated by the functionality of the target organ; in fact, we have shown that polyneuronal innervation and sprouting increase daily to reach the maximum level at the 25th day, then they decline. At day 25 the biophysical properties of the muscle fibre membrane are back to normality; this must trigger the release of some unknown factors which induce the retraction of the redundant innervation [9, 10]. On the other hand, another signal must indicate to the cell body to stop elongation as soon as the synaptic contact is formed. In fact, the regenerating axons grow for tens of millimeters in two weeks from the crush site to the end plate, but here growth ceases immediately and it takes 6 weeks further to regenerate a morphologically normal nerve ending which is only few hundreds of microns long.

Restoration of Neurotransmitter Release

Neurotransmitter release mechanisms are quite complex and must satisfy the need for synaptic transmission, which is made up of phasic and tonic processes of transmitter secretion. At the neuromuscular junction Ach is contained in vesicular and cytoplasmic pools, the former being responsible for quantal and the latter for molecular release [7, 11]. When an action potential reaches the nerve endings there is an influx of Ca^{++} and a consequent massive release of Ach quanta that determines a localized depolarization of the muscle membrane, named end plate potential. If the depolarization is sufficiently large it will induce an action potential with consequent muscle contraction. The tonic release is represented both by molecules of Ach, which leak from the nerve terminal and never reach the postsynaptic membrane, and by packets of Ach probably contained in presynaptic vesicles, which originate the miniature end-plate potentials (mepps) after collision with the postsynaptic receptors [11, 14]. The frequency of mepps can be increased in a controlled manner by raising the $[K^+]$ or the tonicity of the extracellular fluids. K^+-rich solutions stimulate release in a Ca^{++}-dependent manner perhaps by raising its intracellular levels in a way comparable to the action potential [4]. Conversely, hypertonic solutions stimulate release independently from the presence of Ca^{++} in the bathing medium; and its mode of action seems to be related to rearrangement of membrane components and related submembrane cytoskeletal apparatus [12]. Therefore, when we studied the restoration of synaptic transmission in peripheral nerve regeneration we decided to monitor spontaneous and evoked Ach release with K^+ and hypertonic solutions.

The results were very striking and indicated a slow gradual recovery of normal synaptic efficiency. Within 24 h after the synapse is reformed, nerve terminals secrete enough Ach following nerve action potential to induce muscle action potential and contraction, although the nerve endings are very thin and contain few synaptic vesicles. In addition, the total number of quanta contained in the terminal is much reduced at that time, but in 5 days it is back to normal as measured with black widow spider venom [3]. These results are now somewhat corroborated by the morphology, which shows a normal number of vesicles and the beginning of the organization of presynaptic densities. However, in spite of the apparent maturity of the nerve ending, mepps are very rare even several days after synapse formation. For instance, at day 21 from denervation, which is 7 days after reinnervation of the muscle and its related mechanical activity, mepp frequency is about 0.3/s while normal muscles exhibit a frequency of 2.5/s. If we stimulate the same reinnervated preparation with K^+ the frequency rises to 43/s, while with hypertonic solution it goes to 3, as shown in Table 1.

Table 1. Miniature end plate potentials/s

	At rest	K^+	K^+/rest	Hypertonic	H/rest
Normal muscle	2.5	255	100	60	24
Day 21 from denervation	0.3	43	129	3	10

The table shows the mepp frequency at rest and stimulated with 25 mM K^+ and 50% hypertonic solutions (H). In addition, the ratio between stimulated and resting release (K^+/rest; H/rest) is shown

In this table we show that at this stage of reinnervation mepps are rare even if we stimulate with K^+ or hypertonic solutions, however, the ratio shows that the efficiency of the hypertonic solution is still below 50% of normal value.

The normal mepps frequency is achieved at about 40 days from denervation for both resting and K^+-stimulated release, but only at day 60 for the hypertonic solution. Therefore, it is obvious that it takes 14 days for the regenerating axons to regrow several millimeters and reinnervate the muscle, but it takes a further 46 days to reach full synaptic maturity. It is now well demonstrated that the rate limiting step in axonal regeneration is the slow axonal transport. Slowly transported tubulin, actin and neurofilament proteins make, as main components of the cytoskeleton, an essential contribution to the axonal elongation [13]. In particular, the rate of axonal regeneration corresponds to the transport of the fraction named Slow Component b(SCb), which advances at a velocity of 2–5 mm/day, and different rates of elongation correspond to the rate at which SCb is transported. Particularly, in the rat the SCb and the rate of elongation are identical $\cong$ 4 mm/day [16]. However, while nerve regeneration and the efficiency of both K^+- and nerve-evoked release are achieved in 2 weeks, the restoration of synaptic activity proceeds at a much slower rate regarding spontaneous and Ca^{++}-dependent release. This would suggest that essential synaptic components are transported with at least two different rates; the presynaptic elements for Ca^{++}-dependent release would travel with the velocity of SCb, while the other necessary to the spontaneous and Ca^{++}-dependent release would be in a slower tail, perhaps Slow Component a (SCa), which travels at 1 mm/day.

These results are clearly shown in Fig. 5, which represents the recovery of both the acetylcholinesterase (AchE) flow rate and neuromuscular parameters relative to the time of reinnervation. AchE is a membrane bound enzyme and it is transported

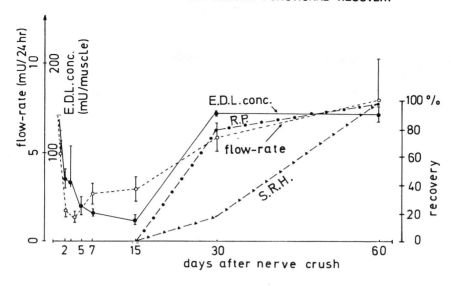

Fig. 5. This figure shows the effects of denervation on AchE flow rate and EDL muscle concentration of AchE. The data are compared to the percentage of recovery of two neuromuscular parameters after reinnervation, which starts at day 15 (recovery). Following crush of the sciatic nerve (day 0) there is a drop in AchE muscle content and flow rate; the minimum value is reached 15 days later for the former, while the latter is already recovering. The muscle membrane resting potential *(R.P.)* linearly recovers after synapse reformation and so also does the AchE concentration in the muscle. However, the recovery of the transmitter release stimulated by hypertonic solutions *(S.R.H.)*, which is only 20% of control values, even 15 days after that synapses have been reformed (day 30), is much slower

with the fast rate at approximately 400 mm/day. Following sciatic nerve crush AchE flow drastically drops and reaches the lowest value at day 4, then begins to recover and a normal rate is obtained between day 30 and 60. It is therefore possible that the signal related to the resumption of AchE transport occurs after axonal sprouting and is related in time to the crossing of the injury site. Perhaps the trigger is the contact of the growing tips with the intact basal lamina of the distal stump.

   These results demonstrate that fast axonal transport of the structures related to AchE flow may carry the components of the Ca^{++}-dependent release mechanism and that the components of the Ca^{++}-dependent release mechanisms are much slower.

   We have also studied the concentration of AchE in the EDL muscles at various time intervals after denervation. The lowest level is reached at day 15, but then muscle reinnervation begins and the concentration linearly recovers within 2 weeks. It is interesting to note that the recovery in AchE concentration is parallel to the recovery of muscle membrane resting potential, suggesting that its regulation is due to the reinnervation of the muscle and consequent mechanical activity. These data confirm the results of neural and muscular influence on AchE published by Lømo and Slater [17]. A detailed description of axonal transport in peripheral nerve regeneration is given by Di Giamberardino et al. in this volume.

The Role of Axoplasmic Transport in Axonal Sprouting

In the previous section I mentioned that perhaps synaptic components are transported at different rates during axonal regeneration and this could explain the well-differentiated synaptic restoration. However, another good possibility is that as soon as the contact between growing axonal tips and denervated end plates is made and the synapse is formed, a block or an extreme slowing down of the anterograde transport takes place at the preterminal part of the axon. This may also imply that the preterminal site of the axon may be the place where reversal of the transport occurs.

This could be of great importance in understanding why ultraterminal sprouts may grow only for a few tens of microns, while collaterals can be much longer. The former are direct elongations of the releasing nerve terminals and the latter are formed at the node of Ranvier [10].

I would like to suggest that if the preterminal site of the axons is the site of transport reversal, ultraterminal sprouts may not grow for long distances because they are not sufficiently supplied by the transport; however, the collateral sprouts formed at the node of Ranvier are directly hooked on the vital stream moving down from the cell body and are able to hold a prolonged growth. This suggestion also raises the question of whether or not ultraterminal sprouts possess a proper cytoskeleton.

Conclusions

Our research suggests that the rate of restoration of synaptic activity may reflect the rate of axonal transport of the various materials and we propose that the Ca^{++}-dependent transmitter release mechanisms are transported with the fast rate regarding membrane bound materials and with the Slow Component b regarding the cytoplasmic components, while the Ca^{++}-dependent transmitter release mechanisms are transported with the Slow Component a.

Another possibility is that transport reversal occurs at the preterminal part of the axons, and therefore the anterograde transport within the terminal is extremely slow regarding, for instance, the cytoplasmic components. This would also suggest that the deficient growth of the ultraterminal sprouts may be due to insufficient supply of cytoskeleton components.

References

1. Axelsson J, Thesleff S (1959) A study of supersensitivity in denervated mammalian skeletal muscle. J Physiol (London) 147:178–193
2. Blinzinger K, Kreutzberg GW (1968) Displacement of synaptic terminals from regenerating motoneurons by microglial cells. Z Zellforsch 85:145–157
3. Carmignoto G, Finesso M, Tredese L, Gorio A (1981) Transmitter release mechanisms during the early stages of reinnervation of a fast twich muscle of the rat. Effects of ganglioside treatments. In: Block K, Bolis L, Tosteson T (eds) Membranes and the environment, responses of membranes to external agents. Raven Press, New York, pp 297–312

4. Cook JD, Okamoto K, Quastel DMJ (1973) The role of the calcium in depolarization secretion coupling at the motor nerve terminal. J Physiol (London) 228:459–497

5. Fambrough DM (1970) Acetylcholine sensitivity of muscle fiber membranes: mechanism of regulation by motoneurons. Science 168:372–373

6. Gorio A (1980) Muscle innervation and reinnervation as a model for development and specificity of neuronal connections. In: Di Benedetta C, Balazs R, Gombos G, Porcellati G (eds) Multi-disciplinary approach to brain development. Elsevier/North Holland Biomedical Press, Amsterdam, pp 439–452

7. Gorio A (1980) Receptors, innervation and neurotransmitter release: microphysiology of chemical synapses. In: Pepeu G, Kuhar MJ, Enna SJ (eds) Advances in biochemical psychopharmacology. Raven Press, New York, pp 57–65

8. Gorio A, Garmignoto G (1981) Reformation, maturation and stabilization of neuromuscular juctions in peripheral nerve regeneration. The possible role of exogenous gangliosides on determining motoneuron sprouting. In: Gorio A, Millesi H, Mingrino S (eds) Post-traumatic peripheral nerve regeneration. Raven Press, New York, pp 481–493

9. Gorio A, Carmignoto G, Facci L, Finesso M (1980) Motor nerve sprouting induced by ganglioside treatment. Possible implications for gangliosides on neuronal growth. Brain Res 197: 236–241

10. Gorio A, Carmignoto G, Ferrari G (1981) Axonal sprouting stimulated by gangliosides. A new model for elongation and sprouting. In: Gorio A, Rapport (eds) Gangliosides in neurological and neuromuscular function, development and repair. Raven Press, New York, pp 177–195

11. Gorio A, Hurlbut WP, Ceccarelli B (1978) Acetylcholine compartments in mouse diaphragm: comparison of the effects of blck widow spider venom, electrical stimulation and high concentration or potassium. J Cell Biol 78:716–733

12. Gorio A, Mauro A (1979) Reversibility and mode of action of black widow spider venom on the vertebrate neuromuscular junction. J Gen Physiol 73:245–263

13. Hoffman PN, Lasek RJ (1980) Axonal transport of the cytoskeleton in regenerating motor neurons: constancy and change. Brain Res 202:317–333

14. Katz B, Miledi R (1975) Transmitter leakage from motor nerve endings. Proc R Soc London Ser B 196:59–72

15. Kreutzberg GW (1981) The regeneration program of the neuron: an introduction. In: Gorio A, Millesi H, Mingrino S (eds) Post-traumatic nerve regeneration. Raven Press, New York, pp 3–6

16. Lasek RJ, McQuarrie IG, Wujek JR (1981) The central nervous system regeneration problem: neuron and environment. In: Gorio A, Millesi H, Mingrino S (eds) Post-traumatic nerve regeneration. Raven Press, New York, pp 59–74

17. Lφmo T, Slater CR (1980) Control of junctional acetylcholinesterase by neural and muscular influences in the rat. J Physiol (London) 303:191–202

18. Mark RF (1980) Synaptic repression at neuromuscular junctions. Physiol Rev 60:355–395

19. Miledi R (1960) The acetylcholine sensitivity of frog muscle fibers after complete or partial denervation. J Physiol (London) 151:1–23

20. Sanes JR, Marshall LM, MacMahan UJ (1978) Reinnervation of muscle fiber basal lamina after removal of myofibers. Differentiation of regenerating axons at original synaptic sites. J Cell Biol 78:176–198

21. Sperry RW (1950) Myotypic specificity in teleost motoneurons. J Comp Neurol 93:277–287

22. Sperry RW (1963) Chemoaffinity in the orderly growth of nerve fiber patterns and connections. Proc Natl Acad Sci USA 50:703–710

Section 3　Experimental Neuropathies and Axoplasmic Transport

Disruption of Axoplasmic Transport by Neurotoxic Agents. The 2,5-Hexanedione Model

PETER S. SPENCER [1] and JOHN W. GRIFFIN [2]

Introduction

Toxic compounds which interfere with axonal integrity and transport are many and varied. Among the first to be utilized were the spindle inhibitors such as colchicine and the vinca alkaloids (e.g. vincristine, vinblastine). These compounds, together with podophyllotoxin and the maytansinoids (maytansine and maytanprine), possess a special affinity for tubulin, block the formation of axonal microtubules and, perhaps for these reasons, disrupt axonal transport. They also induce an accumulation of 10-nm neurofilaments following direct application to nervous tissue, although the basis for this effect is unknown (reviewed in [5]). Systemic administration of such agents (e.g. vincristine) produces a distal, sensori-motor neuropathy in man [2].
A similar type of neuropathy also accompanies systemic treatment with agents which have no known adverse effects on microtubules. Some of these compounds, such as pyridinethione and p-bromophenylacetylurea, induce an abnormality in fast antero-grade transport and/or retrograde return of fast transported materials, as well as an accumulation of tubulo-vesicular profiles, prior to degeneration of distal axons [10]. Other agents apparently inactive on microtubules, such as the hexacarbons (e.g. 2-hexanone, 2,5-hexanedione, 2,4-hexanediol), acrylamide and carbon disulfide, induce early distal axonal changes in the form of multifocal accumulations of 10-nm neurofilaments, together with varying amounts of smooth endoplasmic reticulum, vesicles and mitochondria [17, 21]. A similar pattern of neurofila mentous giant axonal degeneration is seen in proximal axons of animals systemically intoxicated with β,β'-iminodipropionitrile (IDPN), an agent which impairs anterograde transport of neurofilament triplet proteins while producing little distal nerve-fiber breakdown [6]. Some of these agents (2,5-hexanedione, IDPN, acrylamide) induce similar changes in the axonal cytoskeleton following *direct* application to peripheral nerves, and are useful in exploring the relationship between perturbations of axonal transport and the onset of nerve-fiber breakdown. This paper describes recent advances in our understanding of these questions which have come from the use of 2,5-hexanedione (2,5-HD) as an experimental probe.

---

1 Institute of Neurotoxicology, Departments of Neuroscience and Pathology, Albert Einstein College of Medicine, Bronx, NY 10461, USA
2 Neurology, Traylor Building, The Johns Hopkins University School of Medicine, Baltimore, MD 21205, USA

Axoplasmic Transport in Physiology and Pathology
(ed. by D.G. Weiss and A. Gorio)
© Springer-Verlag Berlin Heidelberg 1982

Properties of 2,5-Hexanedione

2,5-HD (syn. acetonyl acetone) is a symmetrical gamma diketone, soluble in water, solvents and oils. The purified compound is inexpensive and readily obtainable. It is used commercially as an intermediate in the synthesis of perfume ingredients, rubber accelerators, dyes, inhibitors, insecticides and pharmaceuticals, as a solvent for inks and wood stains, as a gasoline additive, and as a tanning agent. The neurotoxic properties of 2,5-HD were discovered when the compound was identified as the endogenous metabolite responsible for the induction of peripheral neuropathy in individuals chronically exposed to n-hexane or 2-hexanone. 2,5-HD owes its neurotoxic property to the 1,4-spacing of the carbonyl groups, longer-chain aliphatic gamma diketones (e.g. 2,5-heptanedione, 3,6-ocatanedione) also inducing identical types of neurofilamentous giant axonal swelling. Molecular analogues of 2,5-HD which possess a different spacing of the carbonyl groups, such as 2,4-hexanedione (2,4-HD), are free of chronic neurotoxic properties and serve as valuable negative control compounds in experiments utilizing 2,5-HD [23]. The biochemical mechanism of axonal toxicity is unknown, but 2,5-HD reacts with amino-acid residues to form a Schiff base and a pyrrole, inhibits the glycolytic enzymes glyceraldehyde-3-phosphate dehydrogenase and phosphofructokinase (but not lactic dehydrogenase, transketolase, succinic dehydrogenase or pyruvate decarboxylase), and reduces ATP levels in nerve which can be restored with pyruvate ([15] and Sabri, unpublished data). These observations have led to the suggestion that 2,5-HD impedes axonal transport by blocking glycolysis on which it is known to depend [13, 14, 20]. Glycolytic enzymes are intrinsic components of axoplasm (squid) and at least three (neuron-specific enolase, aldolase, and pyruvate kinase) move along the axon (guinea-pig optic nerve) as part of slow component b [3].

Effects of Systemic Intoxication with 2,5-HD

Animals exhibit the neurotoxic effects of 2,5-HD after prolonged systemic exposure. Rats drinking 0.5% 2,5-HD ad libitum (350–520 mg/kg/day), or the hydroxylated derivative 2,5-hexanediol (2,5-HDiol), which is metabolized to 2,5-HD, develop after 6–8 weeks a slowly progressive hindlimb weakness attributable to distal axonal degeneration and denervation of muscle [21]. Distal axonopathy and muscle denervation also develop in organotypic cord-ganglion-muscle combination cultures treated with 2.8 mM 2,5-HD or 2,5-HDiol for similar periods of time [25]. Inspection of the distal regions of affected nerve fibers from animals and cultures at stages during intoxication reveals a stereotyped spatial-temporal sequence of pathological change.

The earliest changes in myelinated fibers consist of axonal swellings which develop on the proximal sides of distal nodes of Ranvier (Fig. 1), sometimes at the distal extremity of the unbranched portion of the motor axon [26]. The swollen axon often displaces the paranodal myelin sheath. Ultrastructural examination reveals that the axonal change begins by focal condensation of neurotubules, mitochondria and smooth endoplasmic reticulum (SER) to form a central channel of axoplasm (Figs. 2 and 3), coupled with a massive local increase in the number of 10-nm neurofilaments which, together with scattered mitochondria, often adopt a circumferential orientation. Within a few weeks of the appearance of the proximal-side paranodal swelling in vitro, the adjacent distal-side paranode sometimes becomes enlarged. Distal-side paranodal swellings in nerve fibers removed from rats treated with hexacarbons contain few neurofilaments and accumulations of mitochondria, dense membranous bodies and other vesicular organelles. Prolonged intoxication causes giant axonal swellings to develop in more proximal loci both paranodally and internodally, as well as in unmyelinated fibers. Eventually, affected fibers undergo distal breakdown, a giant axonal swelling marking the interface between the proximal preserved portion, and the distal, degenerating length of nerve fiber (Fig. 4). This suggests that the critical factor responsible for distal fiber breakdown is located within the giant axonal swelling close to the node of Raniver (Fig. 5).

Effects of Local Application of 2,5-HD to Nerves

Structural changes in the axonal cytoskeleton are reproduced focally in nerves exposed directly to 2,5-HD, either by intraneural injection or by focal immersion [8, 12]. Injection of 4M 2,5-HD causes a large proportion of axons to develop (within hours) a central channel, enriched in microtubules, SER, mitochondria and other particulate organelles, surrounded by a perimeter of chaotically oriented organelles

Fig. 1. Giant axonal swelling in the proximal-side paranode of the distal region of a single myelin- ▶ ated nerve fiber removed from the tibial nerve of a rat with subclinical 2,5-HD neuropathy. Note the attenuated and corrugated distal internode *(right)*. Figs. 1 and 4 are light micrographs of tissue fixed with glutaraldehyde and osmium tetroxide, dehydrated, and teased apart in epoxy resin. × 150. [21]

Fig. 2. Giant axonal swelling displaying a densely stained central channel. Light micrograph of a one-micrometer-thick cross-section stained with toluidine blue. Figs. 2 and 3 display tissue removed from animals systemically intoxicated with 2-hexanone. × 1410. [21]

Fig. 3. The central channel of a giant axonal swelling is enriched in microtubules, SER and mitochondria. 10-nm neurofilaments surround the channel. Figs. 3, 6 and 7 are transmission electron micrographs of thin epoxy cross sections stained with uranyl acetate and lead citrate. × 13,000. [21]

Fig. 4. Interface between proximal preserved region and distal degenerated portion of the distal region of a single myelinated nerve fiber removed from the tibial nerves of a rat with 2,5-HD neuropathy. × 150. [21]

Fig. 5. Diagram depicts the early paranodal changes in distal regions of nerve fibers in the sciatic nerve of animals systemically intoxicated with neurotoxic hexacarbons such as 2,5-HD. On the proximal side of the node of Ranvier, an abnormally large number of 10-nm neurofilaments accumulates in a chaotic array and causes the axon to swell focally and the myelin sheath to retract away from the node. On the distal side, the paranode has few neurofilaments and may accumulate large numbers of organelles which become sequestered in Schwann-cell cytoplasm (not shown – see text); the nerve fiber is shrunken and the myelin sheath corrugated. The large majority of microtubules, SER and mitochondria (not shown) runs through the paranode as a central channel. [15]

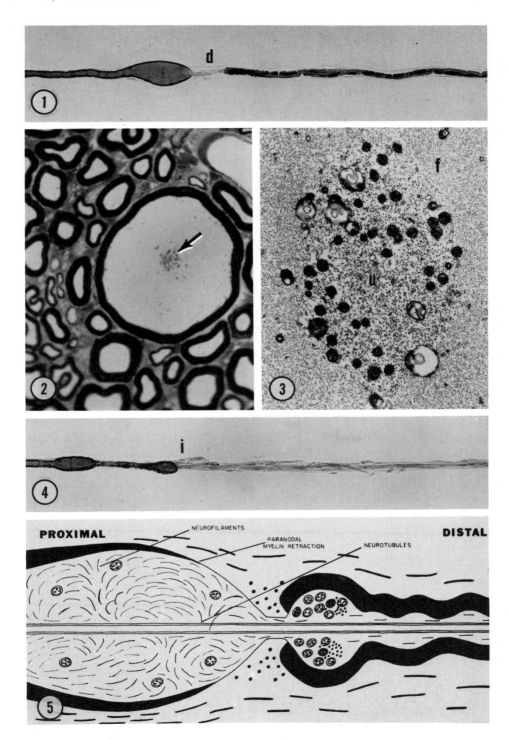

enriched in neurofilaments. Similarly, nerves focally exposed extraneurally for 45 min to concentrated solutions of 2,5-HD display (4–16 days later) a few axons with central channels and many grossly swollen with masses of neurofilaments (Figs. 6, 7). Comparable changes are not observed following treatment with saline or 2,4-HD.

Fate of Axonal Transport

Although direct evidence is lacking, the observed accumulation of neurofilaments on proximal-side paranodes, coupled with the paucity of neurofilaments distally, suggests that 2,5-HD induces a multifocal impairment of the anterograde movement of the neurofilament triplet proteins. The similarities between the giant axonal swellings produced by 2,5-HD and IDPN, where impairment of neurofilament transport is proven, further supports this interpretation. Experimental studies have demonstrated abnormalities in other phases of axonal transport: temporal reduction in the rate of fast anterograde transport (with 2-hexanone which is metabolized to 2,5-HD) and slowing and reduced delivery of materials by retrograde transport (with 2,5-HD) [10, 11, 16]. The latter observation is consistent with the accumulation of membranous structures on the distal-side paranodal axon of affected distal fibers.

The relationship between the abnormalities in neurofilament transport and the slowing of fast transport has been investigated in rats with early neuropathy induced by chronic oral intoxication 2,5-HDiol, which is also metabolized to 2,5-HD [9]. Tritiated leucine or fucose was injected into the ventral horn of spinal segments L4 and L5 and/or into corresponding dorsal root ganglia. Six and 24 h after labeling, animals were perfused with fixatives and the sciatic nerve in the lower thigh prepared for light and electron autoradiography. Control nerves removed from untoxicated animals revealed grains overlying the axon and concentrated at nodes of Ranvier (Fig. 8). By contrast, cross-sections of nerves removed from intoxicated animals exhibited fibers with large numbers of grains amassed in giant axonal swellings (Fig. 9). Label was associated with the centralized channels rich in microtubules and smooth endoplasmic reticulum and, especially after 24 h, distributed along the perimeter of swollen axons beneath the axolemma, presumably in association with organelles such as mitochondria (Fig. 10). When the number of grains in cross-sections

---

Fig. 6. Swollen myelinated axon in a tibial nerve of a rat 16 days after local application of 2,5-HD ▶ to the nerve. The axoplasm consists of a central channel enriched in microtubules, mitochondria and SER, sourrounded by a region containing many neurofilaments arranged circumferentially. × 19,800. [12]

Fig. 7. Swollen, thinly myelinated nerve fiber in a tibial nerve of a rat 16 days after a local application of 2,5-HD to the nerve. The axon is packed with 10-nm neurofilaments which run in various directions. Mitochondria, neurotubules and SER profiles are segregated in small channels which are free of neurofilaments. × 17,150. [12]

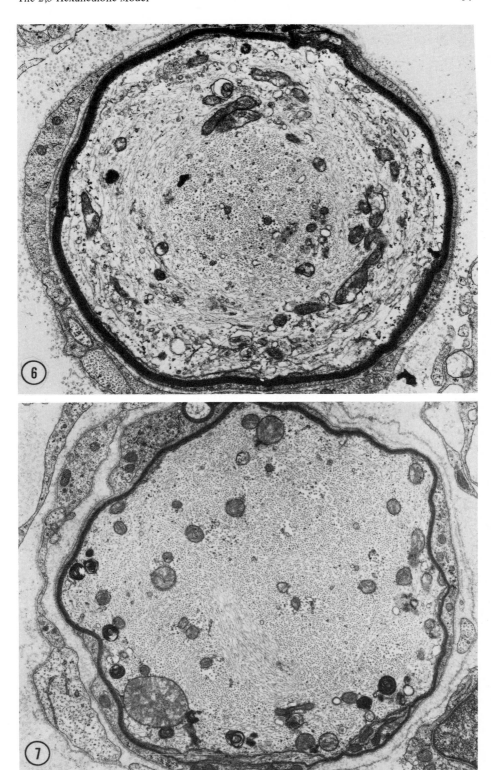

of different levels of individual myelinated fibers was quantified over a length of approximately 300 μm, it was apparent that the grain numbers, and often the grain density, was greater in the neurofilamentous swellings than in other parts of the axon (Fig. 13). Fast transported label also appeared on distal-side paranodes (Fig. 11), but the grain density was strikingly lower than in the adjacent swollen, proximal-side paranodes (Fig. 13). This suggests that fast transported material is held-up during passage across the swollen region, consistent with the observed slowing of the overall rate of fast axonal transport in the affected nerves. This interpretation is strengthened by an absence of change of the rate of fast transport across regions of nerve exposed focally to 2,5-HD prior to the accumulation of 10 nm neurofilaments (Griffin et al., in press).

The Fate of Trapped Organelles

Myelinated nerve fibers seem to possess a mechanism for the sequestration and removal of axonal organelles which are trapped or unable to be transported [24]. Mitochondria, multivesicular bodies, dense membranous bodies, and vesicles with and

---

Fig. 8. Light autoradiogram of a myelinated nerve fiber from the sciatic nerve of a normal rat ▶ demonstrating the differential grain density in the nodal and internodal axoplasm. ^{3}H-fucose was injected into the lumbar ventral horn 24 h before the nerve was fixed. Note the collection of silver grains retained within the constricted segment (myelin sheath attachment sites and node of Ranvier), and the lower densities of grains in the proximal and distal internodes (left and right of the node, respectively). Figs. 8 and 9 are, respectively, longitudinal and epoxy cross sections exposed to Kodak NTB2 for 6 weeks, developed in Kodak D-19 and stained with toluidine blue. × 1,600

Fig. 9. Light autoradiogram of the sciatic nerve from a rat systematically intoxicated with 2,5-HDiol for 6 weeks and perfused with fixatives 6 h after intraspinal injection of ^{3}H-fucose. Note that several axonal swellings contain dense collections of grains, often arranged in subaxolemmal rings (corresponding to the subaxolemmal rings of maloriented neurofilaments). × 850

Fig. 10. Electron-microscope autoradiogram of a transverse section of a neurofilamentous axonal swelling from the sciatic nerve of a rat systemically intoxicated with 2,5-HDiol for 8 weeks and perfused with fixatives 24 h after intraspinal injection of ^{3}H-leucine. This swelling was immediately proximal to a node of Ranvier, as determined by skipserial sectioning. Labeled organelles are present in areas of axoplasm containing microtubules, but are retained in abnormally great amounts in regions containing maloriented neurofilaments. Figs. 10 and 11 are epoxy sections coated with Ilford L4 emulsion, exposed for several weeks, stained en bloc with uranyl acetate and lead citrate. × 45,000

Fig. 11. Electron-microscope autoradiogram from a level distal to the same node of Ranvier as shown in Fig. 10. Note that the fiber is markedly reduced in caliber and a Schwann-cell ingrowth has engulfed axoplasmic vesicular organelles *(right)*. × 34,000

Fig. 12. Short filopodal extension of the axolemma, surrounding invaginations of the adaxonal Schwann-cell cytoplasm, delimit honeycombed structures containing sequestered axoplasm. × 20,000. [4]

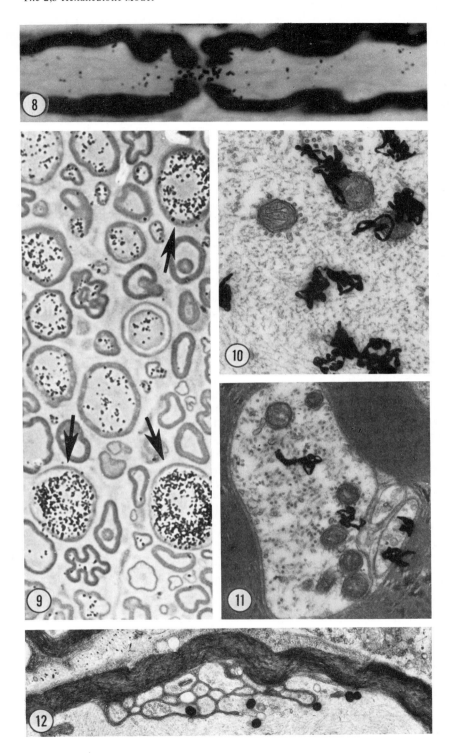

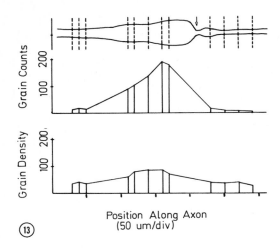

Fig. 13. Distribution, number and density of axonal grains in step cross-sections of six individual myelinated nerve fibers in the sciatic nerve of a rat with 2,5-HDiol neuropathy 24 h after intraspinal injection of ^{3}H-fucose. Note the increased number *(upper graph)* and density *(lower graph)* of grains in the proximal paranodal swelling, and the presence of grains in distal attenuated axons

without dense cores, commonly are sequestered in invaginations of Schwann-cell cytoplasm (Figs. 11, 12). The proposed sequence of events is described in the legend to Fig. 14. The sequestration phenomenon is especially common in the blind axoplasmic pockets of the fluted paranodes of large-diameter myelinated fibers in roots of normal and aged animals [1, 19]. During systemic intoxication with axonal toxins, including 2,5-HD and acrylamide, organelle accumulation is greatly exaggerated and sequestration occurs at multiple paranodal and internodal loci. Although quantitative studies are not available, grains marking the position of fast transported label can be found in autoradiographs overlying sequestered portions of axoplasm containing SER and vacuolar organelles in nerve fibers of animals systemically intoxicated with 2,5-HD (Fig. 11) and acrylamide (Fig. 12) [4, 9, 18]. The sequestration phenomenon may represent a significant method of membrane turnover in the normal axon, and provide a mechanism by which effete and degenerating organelles can be selectively removed without impairing axonal integrity.

Why Does the Axon Degenerate Distally?

The critical event which precipitates axonal breakdown is far from clear. Blockade of neurofilament transport is unlikely to be responsible for two reasons: (a) Results obtained with IDPN demonstrate that distal degeneration need not accompany proximal blockade of neurofilament transport; in most models, distal axonal atrophy is the major alteration (an exception has been found in IDPN-treated cats, which develop distal axonal swellings and fiber loss similar to that produced by 2,5-HD [7]); and (b) neurofilaments are found in the shrunken axons distal to giant axonal swellings in 2,5-HD neuropathy. Complete blockade of fast transport is another possibility, but this has not been demonstrated in 2,5-HD or in the similar experimental neuropathy induced by systemic acrylamide intoxication.

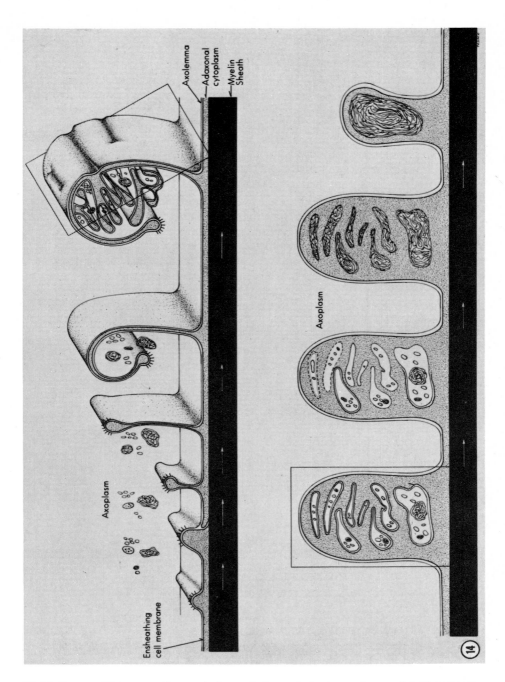

Fi. 14. Process of Schwann-cell sequestration and phagocytosis of axonal organelles. The process starts with the formation of a ridge of adaxonal cytoplasm adjacent to axonal organelles and to an internally coated region of axolemma. The ridge of Schwann-cell cytoplasm enlarges to form a thin sheet which indents the axon surface. The invaginating adaxonal cytoplasmic sheet surrounds the axonal organelles and segregates them from the remainder of the axon. The cytoplasmic sheet infolds on itself *(upper right)* and sequesters groups of axoplasmic organelles to form an inter-digitated or honeycomb structure when viewed in section *(lower, extreme left)*. The two mem-branes (inner axolemma and outer Schwann-cell plasmalemma) surrounding each portion of seques-tered axoplasm become disrupted. Axoplasmic organelles are then taken up by the surrounding Schwann-cell cytoplasm which subsequently retracts to an adaxonal position. [24]

An alternative explanation involves consideration of the fate of the large numbers of degenerating mitochondria and vesicular organelles which accumulate within distal axons, possibly as a result of multifocal blockade of retrograde transport. Although many of these organelles are sequestered and removed from the axon by the phagocytic action of the Schwann cell (vide supra), it is conceivable this mechanism is unable to cope with the large number of accumulating organelles. If these organelles broke down within the axon, their contents would be liberated directly into the axoplasm. Rupture of mitochondria and vesicular organelles would presumably release sequestered calcium into the axoplasm, activate calcium-stimulated protease, and induce local breakdown of axonal integrity [22]. This would be equivalent to transecting the nerve fiber and would precipitate the Wallerian-like distal degeneration seen in 2,5-HD neuropathy. Although this explanation is entirely speculative, it is noteworthy that in the IDPN lesion, a variable number of vesicular organelles accumulate, sequestration of trapped organelles appears less active, and Wallerian-like degeneration is uncommon.

Acknowledgments. P.S.S. was supported by USPHS grants OH 00535 or OH 00851, and J.W.G. by USPHS grants 14784 and 5-POI-NS-10920, and a Research Career Development Award NS-004501. Dr. B. Droz and colleagues kindly granted permission to reproduce Fig. 12 (from [4]).

References

1. Berthold C-H (1978) Morphology of normal peripheral axons. In: Waxman S (ed) Physiology and pathobiology of axons. Raven Press, New York, pp 3–63
2. Bradley WG, Lassman LP, Pearce GW, Walton JN (1970) The neuromyopathy of vincristine in man. J Neurol Sci 10:107–131
3. Brady S, Lasek RJ (1981) Nerve-specific enolase and creatine phosphokinase in axonal transport: soluble proteins and the axoplasmic matrix. Cell 23:515–523
4. Chrétien M, Patey G, Souyri F, Droz B (1981) 'Acrylamide-induced' neuropathy and impairment of axonal transport of proteins. II. Abnormal accumulations of smooth endoplasmic reticulum as sites of focal retention of fast transported proteins. Electron microscope autoradiographic study. Brain Res 205:15–28
5. Ghetti B (1980) Experimental studies on neuronal degeneration. In: Amaducci L, Davison AN, Antuono P (eds) Aging of the brain and dementia, vol 13. Raven Press, New York, pp 183–198
6. Griffin JW, Hoffman PN, Clark AW, Carroll PT, Price DL (1978) Slow axonal transport of neurofilament proteins: impairment by β,β'-iminodipropionitrile. Science 202:633–635
7. Griffin JW, Gold BG, Cork LC, Price DL, Lowndes HE (1982) IDPN neuropathy in the cat: Coexistence of proximal and distal axonal swellings. Neuropathol Appl Neurobiol (in press)
8. Griffin JW, Price DL, Hoffmann PN, Cork LC (1981) The axonal cytoskeleton: Alterations of organization and axonal transport in models of neurofibrillary pathology. J Neuropathol Exp Neurol 40–316
9. Griffin JW, Price DL, Spencer PS (1977) Fast axonal transport through giant axonal swellings in hexacarbon neuropathy. J Neuropathol Exp Neurol 36:603
10. Mendell J, Sahenk Z (1980) Interference of neuronal processing and axoplasmic transport by toxic chemicals. In: Spencer PS, Schaumburg HH (eds) Experimental and clinical neurotoxicology. Williams and Wilkins, Baltimore, pp 139–160
11. Mendell JR, Sahenk Z, Saida K, Weiss HS, Savage G, Couri D (1977) Alterations of fast axoplasmic transport in experimental methyl n-butyl ketone neuropathy. Brain Res 133:107–118

12. Politis MJ, Pellegrino RG, Spencer PS (1980) Ultrastructural studies of the dying-back process. V. Axonal neurofilaments accumulate at sites of 2,5-hexanedione application: evidence for nerve fibre dysfunction in experimental hexacarbon neuropathy. J Neurocytol 9:505–516

13. Sabri MI, Ochs S (1971) Inhibition of glyceraldehyde-3-phosphate dehydrogenase in mammalian nerve by iodoacetic acid. J Neurochem 18:1509–1514

14. Sabri MI, Ochs S (1972) Relation of ATP and creatine phosphate to fast axoplasmic transport in mammalian nerve. J Neurochem 19:2821–2828

15. Sabri MI, Spencer PS (1980) Toxic distal axonopathy: biochemical studies and hypothetical mechanisms. In: Spencer PS, Schaumburg HH (eds) Experimental and clinical neurotoxicology. Williams and Wilkins, Baltimore, pp 206–219

16. Sahenk Z, Mendell JR (1982) Acrylamide and 2,5-Hexanedione neuropathies. Abnormality, of axoplasmic transport in distal axons. Brain Res 219:397–405

17. Seppalainen AM, Haltia M (1980) Carbon disulfide. In: Spencer PS, Schaumburg HH (eds) Experimental and clinical neurotoxicology. Williams and Wilkins, Baltimore, pp 356–373

18. Souyri F, Chretien M, Droz B (1981) 'Acrylamide-induced' neuropathy and impairment of axonal transport of proteins. I. Multifocal retention of fast transported proteins at the periphery of axons as revealed by light microscope autoradiography. Brain Res 105:1–13

19. Spencer PS, Ochoa J (1981) The mammalian peripheral nervous system in old age. In: Johnson J (ed) Aging and cell structure. Plenum Press, New York (in press)

20. Spencer PS, Sabri MI, Schaumburg HH, Moore C (1979) Does a defect in energy metabolism in the nerve fiber cause axonal degeneration in polyneuropathies? Ann Neurol 6:501–507

21. Spencer PS, Schaumburg HH (1977) Ultrastructural studies of the dying-back process. III. The evolution of experimental peripheral giant axonal degeneration. J Neuropathol Exp Neurol 36:276–299

22. Spencer PS, Schaumburg HH (1978) Pathobiology of neurotoxic axonal degeneration. In: Waxman S (ed) Physiology and pathobiology of axons. Raven Press, New York, pp 265–282

23. Spencer PS, Schaumburg HH, Sabri MI, Veronesi BV (1980) The enlarging view of hexacarbon neurotoxicity. CRC Crit Rev Toxicol 7:279–356

24. Spencer PS, Thomas PK (1974) Ultrastructural studies of the dying-back process. II. The sequestration and removal by Schwann cells and oligodendrocytes of organelles from normal and diseased axons. J Neurocytol 3:763–783

25. Veronesi BV, Peterson ER, Spencer PS (1980) Reproduction and analysis of methyl n-butyl ketone neuropathy in organotypic tissue culture. In: Spencer PS, Schaumburg HH (eds) Experimental and clinical neurotoxicology. Williams and Wilkins, Baltimore, pp 863–871

26. Veronesi BV, Peterson ER, Spencer PS (1982/3) Ultrastructural studies of the dying-back process. VI. Examination of nerve fibers undergoing giant axonal degeneration in organotypic culture. J Neuropathol Exp Neurol (in press)

Axoplasmic Transport in the Experimental Neuropathy Induced by Acrylamide

BERNARD DROZ, MONIQUE CHRETIEN, FRANÇOISE SOUYRI,
and GILLES PATEY [1]

Dying back neuropathies are generally characterized by axonal swellings in the distal part of peripheral nerves and by accumulation of neurofilaments in these axonal enlargements (see review [13]). Pioneer investigations on axonal flow of labeled proteins have indicated that acrylamide intoxication interrupted slow axoplasmic transport in peripheral nerves, preferentially in sensory neurons [10]. However, the impairment of slow axoplasmic flow was not confirmed by other studies [1, 3, 14]. On the contrary, various alterations of fast axonal transport were reported in acrylamide-treated animals. The velocity of fast transported proteins was found to be reduced [1, 8]. Local accumulations of membranous tubulovesicular elements were observed in the central nervous system (CNS) [15] and in peripheral nerves [2, 3, 7]. It has been suggested that acrylamide would produce a defect of utilization or removal of fast axonally transported constituents [8, 9]. In acute or chronic intoxication a decreased rate of protein synthesis was found in nervous tissue [11].

To clarify the complexity of the effects of acrylamide on axoplasmic flow, we decided, by means of radioautography, to study the distribution of fast and slowly transported proteins before clearcut lesions appear in distal segments of axons. For this aim, the cholinergic neurons of the preganglionic nerves of the chicken ciliary ganglion were chosen because their 1 cm-long axons constitute a homogeneous population which is affected relatively late by acrylamide. The delayed effects of acrylamide on these short axons allowed us to analyze the distribution of axonally transported proteins when the lesions of the developing axonopathy are still very discrete.

Material and Methods

Forty 3-day old chickens were given 8 I.P. injections of acrylamide at the dose of 100 mg/kg body weight every 2 days. Control chickens were injected with solvent only, under the same conditions. They were sacrified 2 days after the eighth injection. Either 3 h or 7 days before sacrifice, ^{3}H-lysine was delivered into the third cerebral

1 Département de Biologie, Commissariat à l'Energie Atomique, Centre d'Etudes Nucléaires de Saclay, 91191 Gif-sur-Yvette, France

Axoplasmic Transport in Physiology and Pathology
(ed. by D.G. Weiss and A. Gorio)
© Springer-Verlag Berlin Heidelberg 1982

ventricle, that is in the vicinity of the nerve cell bodies of the preganglionic neurons. Ciliary ganglia which enclose the distal part of the preganglionic axons were processed for light and electron microscope radioautography. A quantitative analysis of the label distribution was performed in axons and nerve endings [4, 12].

Results and Discussion

Slowly Transported Proteins

Light microscope radioautography of preganglionic axons performed 7 days after ^{3}H-lysine injection exhibited a concentration of labeled proteins which was either similar or slightly increased in acrylamide treated chickens as compared to controls. Thus, slow axoplasmic transport seemed to be unaffected in 60%, and only slightly altered in 40%, of the acrylamide intoxicated chickens although no obvious sign of axonal degeneration could be detected.

Fast Transported Proteins

Light microscope radioautographs examined 3 h after ^{3}H-lysine injection revealed that about 25% of the axonal population in acrylamide treated chickens showed a threefold enhancement of the relative concentration of labeled protein as compared with controls. Most intriguing was the fact that clusters of silver grains were focally distributed along the preterminal part of axons, whereas about 25% of the axon terminals contained a decreased concentration of label. It could be therefore suspected that fast transported proteins were partly retained in axons and partly delivered to axon terminals. In other words, multifocal accumulations of labeled proteins in axons would reflect a local stasis of fast transported material. Electron microscope radioautography pointed to focal accumulations of both labeled proteins and smooth endoplasmic reticulum tubulo-vesicular profiles and mitochondria (Figs. 1 and 2). Examination of 0.5-1-µm-thick sections previously impregnated with heavy metal salts [16] indicated that the smooth endoplasmic reticulum was locally disorganized. Instead of forming elongated tubules which are running parallel to the long axis of the axon and loosely anastomosed, the smooth endoplasmic reticulum extends focally underneath the axolemma, into an abnormal meshwork of thin interconnected tubules intermingled with mitochondria and small vesicles. Sometimes the disorganized tubular network is encompassed by axolemmal infoldings. Thus the focal stasis of fast transported protein occurs in specific locations characterized by an abnormal configuration of the smooth endoplasmic reticulum (Fig. 3). It may be inferred that the initial stages of acrylamide-induced neuropathy do not reflect a lesion of the motor mechanism of fast axonal transport proper but rather an alteration of the distribution of fast transported membrane constituents [6]. This interpretation is supported by the results of Couraud and Di Giamberardino who found

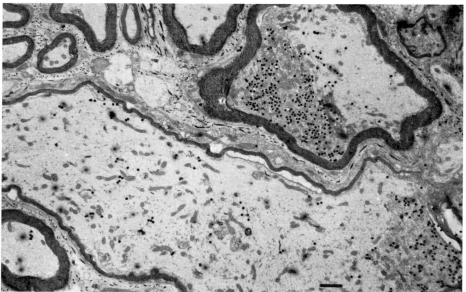

Fig. 1

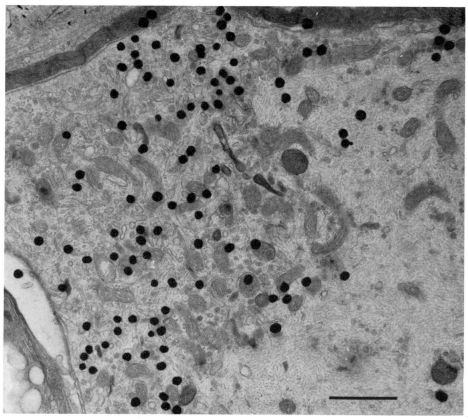

Fig. 2

Figs. 1 and 2. Electron microscope radioautographs of myelinated axons 3 h after injection of ^{3}H-lysine into the cerebral ventricle of acrylamide-treated chickens. In **Fig. 1**, two axons display focal accumulations of silver grains underneath the axolemma. **Fig. 2** shows that the label is associated with mitochondria and numerous tubulo-vesicular profiles of the smooth endoplasmic reticulum [4]. *Bars* 1 μm

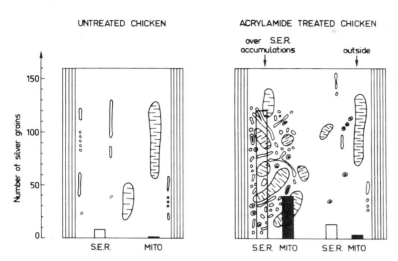

Fig. 3. Diagrammatic representation of the label distribution in axons of untreated or acrylamide-treated chickens. By 3 h after the intracerebral injection of ^{3}H-lysine, quantitative analysis of the radioautographs points to a tremendous increase of the radioactivity in axonal areas occupied by accumulations of the smooth endoplasmic reticulum and mitochondria [6]. *Vertical axis* represents number of silver grains/100 μm²

a partial immobilization of the 20 S molecular form of acetylcholinesterase in acrylamide-treated chickens, whereas the 20 S form is entirely mobile in control animals [5, 7, 17].

In conclusion, the local stasis of fast transported membranous constituents at the distal part of axons may disturb the distribution of axolemmal and presynaptic macromolecules in initiating dying back neuropathies.

References

1. Bradley WG, Williams MH (1973) Axoplasmic flow in axonal neuropathies. Brain 96:115–120
2. Cavanagh JB, Gysbers MF (1980) "Dying back" above a nerve ligature produced by acrylamide. Acta Neuropathol 51:169–177
3. Chretien M, Souyri F, Patey G (1979) Altération du transport axonal rapide et anomalie concomitante du reticulum endoplasmique lisse dans la neuropathie à l'acrylamide. CR Acad Sci (Paris) Sér D 288:689–692

4. Chretien M, Patey G, Souyri F, Droz B (1981) "Acrylamide-induced" neuropathy and impairment of axonal transport of proteins. II. Abnormal accumulations of smooth endoplasmic reticulum at sites of focal retention of fast transported proteins. Electron microscope radioautographic study. Brain Res 205:15–28

5. Di Giamberardino L, Couraud JY, Chretien M, Souyri F (1978) Acrylamide induced experimental neuropathy: effect on the axonal transport of acetylcholinesterase molecular forms in chicken sciatic nerve. Neurosci Lett Suppl 1:11

6. Droz B, Chretien M (1980) Axonal flow and toxic neuropathies. In: Holmstedt B, Lauwerys R, Mercier M, Roberfroid M (eds) Mechanisms of toxicity and hazard evaluation. Elsevier/North-Holland Biomedical Press, Amsterdam New York, pp 13–25

7. Droz B, Koenig HL, Di Giamberardino L, Couraud JY, Chretien M, Souyri F (1979) The importance of axonal transport and endoplasmic reticulum in the function of cholinergic synapse in normal and pathological conditions. In: Tucek S (ed) The cholinergic synapse. Progress in brain research, vol 49. Elsevier, Amsterdam, pp 23–44

8. Griffin JW, Price DL, Drachman DB (1977) Impaired axonal regeneration in acrylamide intoxication. J Neurobiol 8:355–370

9. Griffin JW, Price DL, Engel WK, Drachman DB (1977) The pathogenesis of reactive axonal swellings: role of axonal transport. J Neuropathol Exp Neurol 36:214–227

10. Pleasure DE, Mishler CK, Engel WK (1969) Axonal transport of proteins in experimental neuropathies. Science 166:524–525

11. Schotman P, Gipon L, Jennekens FG, Gispen WH (1977) Polyneuropathies and C.N.S. protein metabolism. II. Changes in the incorporation rate of leucine during acrylamide intoxication. Neuropathol Appl Neurol 3:125–136

12. Souyri F, Chretien M, Droz B (1981) "Acrylamide-induced" neuropathy and impairment of axonal transport of proteins. I. Multifocal retention of fast transported proteins at the periphery of axons as revealed by light microscope radioautography. Brain Res 205:1–13

13. Spencer PS, Schaumburg HH (1980) Recent morphological studies of toxic neuropathy. In: Holmstedt B, Lauwerys R, Mercier M, Roberfroud M (eds) Mechanisms of toxicity and hazard evaluation. Elsevier/North Holland Biomedical Press, Amsterdam New York, pp 3–11

14. Sumner A, Pleasure D, Ciesielka K (1976) Slowing of fast axoplasmic transport in acrylamide neuropathy. J Neuropathol Exp Neurol (Abstr) 35:319

15. Suzuki K, Pfaff LD (1973) Acrylamide neuropathy in rats, electron microscopy study of degeneration and regeneration. Acta Neuropathol 24:197–213

16. Thiery G, Rambourg A (1976) A new staining technique for studying thick sections in the electron microscope. J Microsc Biol Cell 26:103–106

17. Couraud JY, Di Giamberardino L, Chrétien M, Souyri F, Fardeau M (1982) Acrylamide neuropathy and changes in the axonal transport and muscular content of the molecular forms of acetylcholinesterase. Muscle and Nerve 5:302–312

Axonal Transport in β,β′-Iminodipropionitrile Neuropathy

JOHN W. GRIFFIN [1], PAUL N. HOFFMAN [2], and DONALD L. PRICE [3]

Introduction

The concept that chronic axonal diseases might be caused by a failure of delivery of essential materials from the nerve cell body to the axon has been implicit in neuro-pathological writings since the time of Waller [21]. Following the initial demonstration of axoplasmic transport by Weiss and Hiscoe [22] and the subsequent rapid advances in methodology, the identification of major transport defects was anticipated in a variety of neurological conditions. However, many of the pioneering studies using disease models produced ambiguous results, with abnormalities which were minor, inconsistent, or present only late in the course of the disease. In the past few years, this confusion has begun to be dispelled by detailed, systematic studies of specific experimental models. As the number of studies correlating abnormalities of axonal transport and neuropathologic changes has increased, three provisional conclusions appear warranted:

1. Important defects may be subtle and may only be demonstrated by very specific assays (e.g. "turn-around" studies, ultrastructural correlation, gel fluorography, or EM autoradiography). It is unlikely that sustained, complete interruption of axonal transport will be found in the early stages of slowly evolving disorders; instead, transport defects of such severity are likely to result in more acute axonal degeneration.

2. The demonstration of an axonal transport defect does not necessarily establish a causal basis for the neuropathological condition; rather, some transport defects may be the consequence of axonal pathology (see Spencer and Griffin, this volume).

3. Specific patterns of transport abnormalities may correlate with particular types of structural alterations. For example, the defects in fast and/or retrograde transport in the distal axon produced by agents such as p-bromophenylacetylurea (BPAU) (see Brimijoin et al., this volume) and zinc pyridinethione [16], probably underlie

---

1 Neuromuscular Laboratory, Department of Neurology
2 Department of Ophthalmology
3 Neuropathology Laboratory, Department of Neurology and Pathology; The Johns Hopkins University, School of Medicine, Baltimore, MY 21205, USA

Axoplasmic Transport in Physiology and Pathology
(ed. by D.G. Weiss and A. Gorio)
© Springer-Verlag Berlin Heidelberg 1982

the accumulation of vesicular and tubular organelles in the affected regions. In contrast, β,β'-iminodipropionitrile (IDPN) produces predominantly neurofilamentous pathology, and the primary defect appears to be in the maintenance of normal cytoskeletal organization and the transport of neurofilament proteins.

The Neurotoxicity of IDPN

This chapter will review the current understanding of the structural and functional abnormalities of IDPN intoxication in the peripheral nervous system (PNS). IDPN ($NH[CH_2CH_2CN]_2$) has been of interest to neuroscientists since the recognition 30 years ago of behavioral abnormalities (for review, see [5]). Chou and Hartmann [1, 2] identified and characterized the basic neuropathology of IDPN intoxication in the rat, showing that the proximal axons of large fibers throughout the nervous system accumulated neurofilaments and underwent massive swelling. They recognized the similarity of these changes to those found proximal to nerve constriction by Weiss and Hiscoe [22], and they suggested that a defect in axonal transport might be important in the pathogenesis of this disorder. This possibility has been confirmed in several recent studies [5, 8, 11, 23]. In this review, the induction of IDPN neurotoxicity will be described, the morphological consequences of both systemic administration and local intraneural injection summarized, and the changes in axonal transport, both slow and fast, discussed.

Induction of the IDPN Lesion

The IDPN lesion can be produced by several routes of administration and by a variety of dosage schedules. Three preparations are referred to in this presentation:

1. Short-term, high-dose systemic administration. As initially described by Chou and Hartmann [1], rats are given intraperitoneal injections of up to 2.5 mg/kg, a dose best tolerated by small rats (less than 8 weeks of age). This schedule produces an increase in activity, circling and rotatory movements of the head, and a tendency to spontaneous arterial dissection and hemorrhages, due to inhibition of collagen cross-linking (see [5, 19] for discussion).

2. Continuous low-dose schedule. This protocol produces much more gradual development of neuropathological changes without inhibiting growth or weight gain and with only mild behavioral changes and tendency to hemorrhage. Animals can be maintained for long periods of time, allowing studies of the effects of sustained abnormalities on the evolution of the lesion [3, 6, 7, 18].

3. Local intraneural injection. Small volumes of concentrated IDPN solutions can be injected into the endoneurial space of the rat sciatic nerve using micropipettes. The resulting lesion is focal and can be used to test the direct effects of IDPN or its analogs on the axon and to study the early and evolving structural and functional abnormalities produced by these agents [10].

Structural Changes — Systemic Administration

Within 24 h following high-dose administration of IDPN, neurofilaments increase in density within the proximal regions of large axons. In motor axons, which have been most thoroughly studied, early axonal swelling is apparent by 48 h. The swelling begins and affects most severely the intraparenchymal internodes with relative sparing of the initial segment and nodes of Ranvier (Fig. 1). Along a given axon, the first three to five internodes usually appear to undergo swelling. By 2–10 days, large swellings are present within the spinal cord and extend into the ventral root exit zone and a few millimeters into the ventral root. By 14 to 30 days, some axons in the ventral horn have developed spherical intraspinal balloons (up to 150 μm in diameter); these structures are often the result of enormous distention of the first or second internode. At this stage, the swellings are filled with interlaced fascicles of maloriented neurofilaments with a varying number of particulate organelles entrapped within the neurofilamentous network [2, 3, 7]. A substantial number of axons show a central channel of longitudinally-oriented axoplasm, containing a high proportion of microtubules surrounded by a ring of spiralling disoriented neurofilaments [7]. Immunocytochemical methods, using polyclonal antibodies directed against one of the neurofilament triplet proteins (68 kD), show that this pattern — a subaxolemmal ring of neurofilaments around the central channel of microtubules — can be demonstrated all along the nerve fibers [17].

In continuous long-term intoxication, the fibers distal to the swellings undergo a slow reduction in axon caliber [3]. Both the proximal swellings and the distal atrophy are reversible. Within 60 days after cessation of IDPN administration, the swellings within the proximal ventral roots are reduced in size and number; in part, this is the result of distal migration of the accumulated axoplasm [1, 6, 20].

This sequence of proximal axonal swelling and distal axonal atrophy, with recovery by transport of axoplasm down the nerve fiber, mimics in many ways the effects of

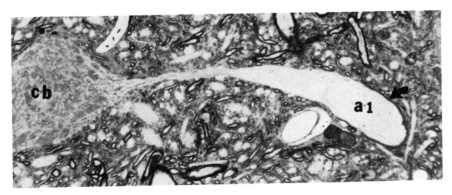

Fig. 1. Lumbar motor neuron from an IDPN-intoxicated rat. A normal nerve cell body *(cb)* and initial segment give rise to an enlarged first heminode of the axon *(al)*. Such axonal swellings contain increased numbers of neurofilaments. Epoxy section stained with toluidine blue. × 670. (From [3] by permission of The Journal of Neuropathology and Experimental Neurology)

focal mechanical constriction of the nerve as initially described by Weiss and Hiscoe [22]. In the latter paradigm, the axoplasm retained proximal to the constrictions is comprised of a large proportion of neurofilaments; microtubules pass through the constriction into distal atrophic segments [4]. The effects of such nerve constrictions are greatest on large caliber axons, a feature this model shares with IDPN. In the case of IDPN, it shoud be emphasized that the primary effect is clearly toxic and cannot be due solely to mechanical or pressure differentials. In the IDPN model, the proximal internodes of large fibers are involved in many parts of the central nervous system; in addition, in some preparations, similar swellings can be established along the course of the nerve fiber [9, 10].

Local Administration Along the Nerve

Recently, we have examined the local effects of concentrated IDPN solutions injected directly into the endoneurial space of rat sciatic nerves as described above [10]. These studies were undertaken in order to demonstrate conclusively that IDPN directly affects the axon, and that development of the pathology does not require participation of the nerve cell body. When a small volume of concentrated IDPN solution was injected beneath the perineurium, the axonal cytoskeleton was dramatically altered, whereas no axonal changes were found when the control solutions were injected into the nerve. Six hours after IDPN injection, light microscopy disclosed a darkly-stained central core surrounded by a glassy, pale peripheral ring (Fig. 2). Electron microscopy disclosed that this target-like appearance was due to a compartmentalization of cytoplasmic organelles [10]. Microtubules collected into a discrete longitudinal channel usually located in the central or paracentral part of the axon. This began 30–60 min after injection and was fully developed by 6 h. Neurofilaments were virtually excluded from the central channels, forming a concentric subaxolemmal ring. At very early times (1/2–2 h) neurofilaments maintained a normal longitudinal orientation but, by 6 h, they had become chaotically arranged. At this stage, most particulate axonal organelles, including mitochondria, membrane-bound vesicles, and dense bodies, segregated with the microtubules in the central channel; the localization of these organelles resulted in the darkly-stained targets visualized by light microscopy. In most fibers, there was no unusual accumulation of organelles in these channels, although occasionally focal collections were seen, particularly near nodes of Ranvier. Even in fibers in which almost all microtubules were segregated into the central channel, some membrane-bound tubules, presumably representing smooth endoplasmic reticulum, remained within the subaxolemmal region.

This segregation of cytoskeletal constituents appeared to be reversible. Within 24–72 h, the axoplasm at the site of injection had assumed a more normal appearance in many fibers. However, in some fibers, the number of neurofilaments was increased and neurofibrillary swellings developed in these regions. In these fibers, local injection reproduced the major features of the proximal swellings occurring after systemic administration. Whorled, interlaced fascicles of neurofilaments appeared to entrap and orient particulate organelles.

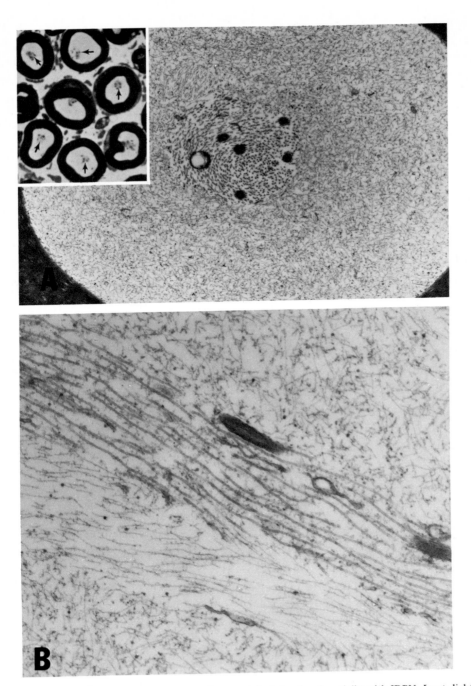

Fig. 2. A Transverse sections of rat sciatic nerve injected subperineurially with IDPN. *Inset:* light micrograph demonstrating central or paracentral collections of particulate organelles *(arrows).* 1 µm epoxy section stained with toluidine blue. The electron micrograph shows that these central channels contain collections of microtubules and particulate organelles, including mitochondria. Surrounding this channel is a ring of chaotically-arranged neurofilaments. Magnification × 10,500 **B** Electron micrograph of a longitudinally sectioned nerve fiber 6 h after local injection of IDPN. The central channel containing microtubules is sharply demarcated from the disoriented neurofilaments. Note a fascicle of longitudinally directed neurofilaments persists beneath the microtubule channel. Magnification × 80,000

Thus, these studies demonstrated a direct effect of IDPN on the axon. The earliest observed change was the disorganization of the axonal cytoskeleton with segregation of microtubules from neurofilaments. These structural findings correlated well with the IDPN-induced abnormalities of axonal transport described below.

Axonal Transport Defects in IDPN Intoxication — Slow Transport

Impairment of the slow axonal transport of neurofilaments is the most prominent abnormality produced by IDPN administration. The initial studies in rat sciatic motor fibers showed that the main slow component peak (SCa of Hoffman and Lasek [13]) was markedly retarded as it passed down the nerve. Polyacrylamide gel fluorography demonstrated that the transport of the neurofilament triplet proteins was severely impaired [8]. However, transport of the 55 kD protein (representing, at least in part, tubulin) as well as actin and other SCb proteins was also abnormal. The extent of abnormalities in tubulin and the SCb proteins varied in part with the specific experimental protocol. In animals given IDPN just before labeling, the abnormalities in transport of these proteins was subtle. However, in chronically intoxicated animals, in whom proximal axonal pathology was well developed before isotopic labeling, movement of the 55 kD protein and actin, as well as the neurofilament triplet proteins, was markedly abnormal [5] (Fig. 3).

These observations suggested that the impairment in transport of the 55 kD protein and actin might be a consequence of a primary defect in neurofilament triplet transport and the consequent neurofilamentous pathology. We asked whether it might be possible to dissociate more completely neurofilament transport from transport of the other slow component constituents. These studies took advantage of the observation that both slow components (SCa and SCb) are transported more than twice as fast in 3-week old animals than in 16-week old animals [14]. For example, SCa has a rate of 2–4 mm/day in 3-week old animals compared to 1 mm/day in young adult animals. When IDPN was administered to 3-week old animals, the rate of neurofilament triplet transport was severely slowed, with large amounts of neurofilament proteins retained in the very proximal portion of the ventral root (Fig. 4). The distribution of the 55 kD protein and of actin along the nerve was measured by cutting out these bands in every track (each track representing a 3 mm nerve segment). The resulting plots of the amount of radioactivity in these proteins along the nerve correspond very nearly to plots obtained from normal animals; there was only minor retention of tubulin in the proximal axonal regions. This preparation allowed almost complete dissociation of the IDPN effect on neurofilament transport from transport of tubulin and actin.

The conclusion that IDPN produces a primary effect on neurofilament transport is supported by recent studies of Yokoyama et al. [23]. In rat sciatic motor and sensory neurons, these investigators confirmed the defect in neurofilament transport described above; inspection of their published data shows some retention of tubulin as well as neurofilaments in proximal nerve regions. However, they performed the important additional experiment of labeling a pure C-fiber population. These small, unmyelinated fibers have very few neurofilaments ultrastructurally, and no neurofilament triplet proteins are recognized in the fluorographic profiles of the slow component. In these neurofilament-poor fibers, they found no abnormality of slow transport.

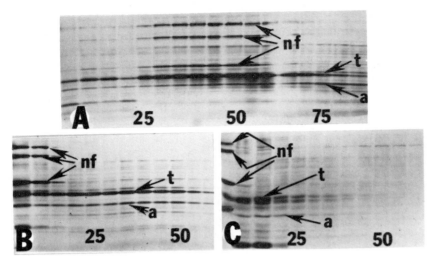

Fig. 3. Fluorograms of polyacrylamide slab gels, showing the distribution of individual labeled proteins along the nerves. Each track represents a 5 mm nerve segment, with the most proximal segment at the far left; the *numbers* (25, 50, or 75 mm) identify the distance of the segments along the nerve. A Contraol animal 21 days after labeling with ^{35}S-methionine. The neurofilament triplet proteins *(nf)* (68, 145, and 200 kD) are present primarily in tracks corresponding to 25 to 50 mm (that is, they have moved 1 to 2.5 mm/day). Tubulin *(t)* and actin *(a)* are present over a wider spectrum of velocities (0.5 to 5 mm/day). B Experimental animal 21 days after intraspinal injection of ^{35}S-Methionine and 20 days after injection of IDPN. The neurofilament triplet proteins are in large part retained in the initial two segments; transport of tubulin and actin is impaired as well, compared to control. C Experimental animal given IDPN in drinking water, 21 days after labeling with ^{3}H-leucine. The marked impairment in slow transport is apparent. (From [8] by permission of the American Association for the Advancement of Science)

These kinetic observations correlate well with the recent morphological studies described above. Systemic administration produces channels containing microtubules, surrounded by neurofilamentous axoplasmic pads [7, 17]. Local administration produces more complete dissociation of neurofilaments from microtubules. These observations suggest that a primary effect of IDPN might be to alter neurofilament-microtubule interactions. The basis for such an effect is unknown. The possibilities include: an excessive cross-linking of neurofilaments to each other; an interruption of neurofilament/microtubule interactions, perhaps mediated through cross-bridging; and a defect in energy metabolism (the last suggested because some aspects of the IDPN-induced pathology resembled those produced by local cyanide administration [12]).

Fast Axonal Transport

The relative normality of fast axonal transport as compared to slow transport is a striking aspect of the neurotoxicity of IDPN. Even in continuously intoxicated animals with severely swollen proximal nerve fibers, the rate of fast transport along the

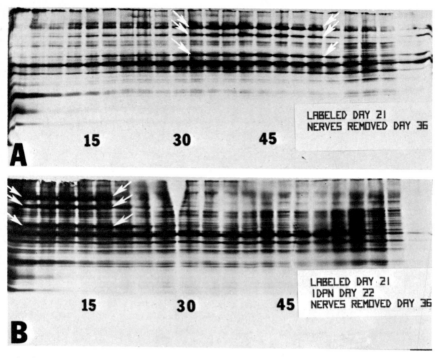

Fig. 4. Fluorograms of polyacrylamide slab gels. Each track represents a 3 mm nerve segment; the *numbers* (15, 30, 45) identify the position of the segments along the nerve. Both fluorograms are from rats injected intraspinally with ^{35}S-methionine on day 21 of life; the nerves were removed 15 days later. The distribution of the largest proportion of the neurofilament triplet proteins is identified by the *white arrows*. Note that in the control animal (**A**) these proteins extend over 50 mm down the nerve. In the IDPN-treated animal (**B**) the movement of the neurofilament proteins is markedly retarded. The other slowly transported proteins are nearly normal in their distribution along the nerve. In studies such as these neurofilament transport is impaired at times when other slow component constituents are unaffected

nerve is normal or nearly normal [5, 11]. However, more subtle abnormalities of fast transport appear to occur following both systemic and local administration of IDPN. The amount of radioactivity in the fast transport peak is reduced compared to comparable groups of normal animals [11]. In addition, the conformation of the transport curves is often altered with a less well-defined peak. Autoradiographic studies show focal collections of rapidly transported organelles within the neurofilamentous swellings, a change closely resembling that described in hexacarbon neuropathy (Spencer and Griffin, this volume). Moreover, Kuzuhara and Chou [5] have recently shown that retrogradely-transported horseradish peroxidase (HRP)-labeled profiles accumulate within giant axonal swellings and that the appearance of HRP in the cell bodies of large motor neurons is delayed. Since large motor neurons have large axons, containing the greatest number of neurofilaments, one might anticipate that these cells, which show the largest swellings, would show the most severe secondary changes in fast anterograde and retrograde transport. Taken together, these observations suggest that proximal axoplasmic disorganization produces a secondary impairment in anterograde and retrograde transport.

The local intraneural administration of IDPN provides a system in which to test the direct effect of this agent on fast transport prior to development of axonal swelling. The movement of the peak of fast transport through an injected segment is nearly normal. However, some fibers show paranodal collections of particulate organelles, and occasionally subaxolemmal accumulations are seen. These findings probably reflect a degree of focal impairment of fast transport in some fibers.

The locally injected nerves, with central microtubule channels and surrounding subaxolemmal neurofilamentous zones, also provide an opportunity to examine the route within the axon utilized by rapidly transported organelles. Electron microscopic autoradiography shows a clear association of rapidly transported materials with the central channels (Griffin and Price, unpublished observation). Interpretation of these results with regard to the mechanism and the necessary structural substrates for transport in normal fibers must be done cautiously, but these observations indicate that fast transport is preferentially, and perhaps necessarily, associated with microtubules. These observations render unlikely models which postulate that axolemmal or subaxolemmal constituents are necessarily involved in the mechanism of fast transport.

Conclusions

The IDPN model provides a simple and reproducible model for examining the interrelationships between axonal transport and neurofibrillary axonal pathology. At present, it appears that the cytoskeletal disorganization and resulting defect in neurofilament transport are the basic abnormalities, and that secondary alterations develop in other aspects of slow and fast transport. The model may provide insight into the pathogenetic mechanism involved in other neurofibrillary axonal degenerations, such as motor neuron disease and some toxic neuropathies.

Acknowledgments. Studies from these laboratories were supported in part by NS 14784, NS 15721, and NS 10580. John W. Griffin is a recipient of an Research Career Development Award NS 004501. Paul N. Hoffman is a John A. and George L. Hartford Fellow (1981–1984), an Alfred P Sloan Research Fellow (1981–1983), and the recipient of a Young Investigator Award (R23 EY 03791). Kenneth Fahnestock provided expert technical assistance in our studies.

References

1. Chou SM, Hartmann HA (1964) Axonal lesions and waltzing syndrome after IDPN administration in rats. Acta Neuropathol 3:428–450
2. Chou SM, Hartmann HA (1965) Electron microscopy of focal neuro-axonal lesions produced by β,β'-iminodipropionitrile (IDPN) in rats. Acta Neuropathol 4:590–603

3. Clark AW, Griffin JW, Price DL (1980) The axonal pathology in chronic IDPN intoxication. J Neuropathol Exp Neurol 39:42–55
4. Friede RL (1971) Changes in microtubules and neurofilaments in constricted, hypoplastic nerve fibers. Acta Neuropathol Suppl V:216–223
5. Griffin JW, Price DL (1980) Proximal axonopathies induced by toxic chemicals. In: Spencer PS, Schaumburg HH (eds) Experimental and clinical neurotoxicology. Williams & Wilkins, Baltimore, pp 161–187
6. Griffin JW, Price DL (1981) Demyelination in experimental IDPN and hexacarbon neuropathies: evidence for an axonal influence. Lab Invest 45:130–141
7. Griffin JW, Price DL (1981) Schwann cell and glial responses in β,β'-iminodipropionitrile intoxication. I. Schwann cell and oligodendrocyte ingrowths. J Neurocytol 10:995–1007
8. Griffin JW, Hoffman PN, Clark AW, Carroll PT, Price DL (1978) Slow axonal transport of neurofilament proteins: impairment by β,β'-iminodipropionitrile administration. Science 202:633–635
9. Griffin JW, Gold BG, Cork LD, Price DL, Lowndes HE (1982) IDPN neuropathy in the cat: coexistence of proximal and distal giant axonal swellings. Neuropathol Appl Neurobiol 8:221–241
10. Griffin JW, Fahnestock KE, Price DL, Hoffman PN (1982) Microtubule neurofilament segregation produced by β,β'-iminodipropionitrile; evidence for the association of fast axonal transport with microtubules. J Neuroscience, in press
11. Griffin JW, Hoffman PN, Fahnestock KE, Price DL (1982) Fast axonal transport in IDPN intoxication. (In preparation)
12. Hall SM (1972) The effects of injection of potassium cyanide into the sciatic nerve of the adult mouse: in vivo and electron microscopic studies. J Neurocytol 1:233–254
13. Hoffman PN, Lasek RJ (1975) The slow component of axonal transport: identification of major structural polypeptides of the axon and their generality among mammalian neurons. J Cell Biol 66:351–366
14. Hoffman PN, Lasek RJ, Griffith JW, Price DL (1982) Slowing of TNG axonal transport of neurofilament protein during development. (Submitted for publication)
15. Kuzuhara S. Chou SM (1981) Retrograde axonal transport of HRP in IDPN-induced axonopathy. J Neuropathol Exp Neurol 40:356
16. Mendell JR, Sahenk Z (1980) Interference of neuronal processing and axoplasmic transport by toxic chemicals. In: Spencer PS, Schaumburg HH (eds) Experimental and clinical neurotoxicology. Williams & Wilkins, Baltimore, pp 139–160
17. Papasozomenos SC, Autilio-Gambetti L, Gambetti P (1982) The IDPN axon: rearrangement of axonal cytoskeleton and organelles following β,β'-iminodipropionitrile (IDPN) intoxication. In: Weiss DG (ed) Axoplasmic transport. Springer, Berlin Heidelberg New York, pp 241–250
18. Parhad IM, Griffin JW, Price DL, Clark AW, Cork LC, Miller NR, Hoffman PN (1982) IDPN intoxication: a toxic model of optic disc swelling. Lab Invest 46:186–195
19. Selye H (1957) Lathyrism. Rev Can Biol 16:1–82
20. Shimono M, Izumi K, Kruoiwa Y (1978) β,β'-iminodipropionitile induced centrifugal segmental demyelination and onion bulb formation. J Neuropathol Exp Neurol 37:375–386
21. Waller AV (1850) Experiments on the section of the glossopharyngeal and hypoglossal nerves of the frog and observations of the alterations produced thereby in the structure of their primitive fibers. Philos Trans R Soc London Ser B 140:423–429
22. Weiss P, Hiscoe HV (1948) Experiments on the mechanism of nerve growth. J Exp Zool 107:315–396
23. Yokoyama K, Tsukita S, Ishikawa H, Kurokawa M (1980) Early changes in the neuronal cytoskeleton caused by β,β'-iminodipropionitrile: selective impairment of neurofilament polypeptides. Biomed Res 1:537–547

Disulfiram-Induced Impairment of Horseradish Peroxidase Retrograde Transport in Rat Sciatic Nerve

ARCHINTO P. ANZIL, GERHARD ISENBERG, and GEORG W. KREUTZBERG[1]

A number of chemical compounds, pathological states and even physiological conditions are known to influence axonal transport [8, 12], and a decreased axonal transport may play a role in the pathogenesis of peripheral neuropathies [3]. We studied the relation between impaired axonal transport and nerve damage using the disulfiram neuropathy model of the rat [1]. This animal model was developed as an experimental counterpart to the neuropathy observed occasionally in humans during prolonged disulfiram (Antabuse ®, Alcophobin ®) treatment for chronic alcoholism [2, 11]. We found a reduction of the conveying capacity of the rat sciatic nerves exposed to disulfiram. Retrograde translocation of horseradish peroxidase (HRP) from the periphery to the corresponding cell bodies was used as indicator of the actual transport taking place in the affected nerves.

Results and Discussion

Albino rats about 300 g of body weight were used in this study. Disulfiram (Espéral, Farmadis, Toulouse, France) in the form of a dry sterile powder was applied to the distal part of the sciatic nerve according to a technique already described [1]. Briefly, the animals were anesthetized by intraperitoneal nembutal ® (40 mg/kg) injection. Working under a minimum of aseptic conditions, the right (experimental) sciatic nerve was exposed at the level of the popliteal fossa. A strip of plastic material was placed around the nerve, filled with disulfiram powder and closed on the top. The same operation was performed on the left sciatic nerve, but the plastic cuff was either left empty or filled with bovine serum albumin powder or autoclaved fine grit. Upon completing the surgical procedure, the muscle layer over the nerve was recomposed and the skin wound closed with surgical clips. About 24 h later the rats were anesthetized and the sciatic nerve was re-exposed and cut about 2 mm distal to the emplacement of the disulfiram or control cuff. Immediately after transection the proximal stump was sucked into a plastic tube. Freshly prepared HRP solution was injected into the tube and delivered to the cut end of the nerve. After 5 min the

1   Max-Planck-Institut für Psychiatrie, Kraepelinstr. 2, D-8000 Munich 40, FRG

the remaining solution was sucked off to terminate HRP uptake. The application of 200–250 mg/HRP (Type IV, Sigma) per ml of phosphate-buffered saline proved to be adequate for the purpose of our experiments. The surgical wound was again closed. About 44 h after HRP application, the rats were anesthetized again and 3 ± 1 mm long experimental and control nerve segments were removed, fixed for 2 h in 2% glutaraldehyde solution at room temperature and kept for 30 addional min in tris-maleate sucrose buffer. Histochemical visualization of HRP was done on longitudinally cut, 30 μm thick, frozen sections treated with diaminobenzidine [7]. Immediately after obtaining the nerve specimens the rats were perfused through the left ventricle with 2.5% glutaraldehyde and spinal cord segments L_2 through L_5 were obtained and placed in additional glutaraldehyde for 2 more h at 4°C. Spinal cord pieces were superficially nicked midway between the ventral median fissure and the dorsal septum in order to discriminate between experimental and control side. 40 μm frozen sections were cut across spinal cord segments followed by visualization of peroxidase with the TMB method [10]. For every spinal cord sample about 50 sections were picked up and examined by light microscopy. Labeled neurons, regardless of their location, were scored + to +++ according to the amount of staining. In addition and independently of the amount of stained material, all labeled neurons in the right hemicord and in the left hemicord were added up accordingly in two separate tallies. An index $I = 1 - n/N$, was calculated of the axonal transport efficiency reduction, where n stood for the total number of labeled neurons in the right hemicord corresponding to the ipsilateral (experimental) sciatic nerve, and N stood for the total number of labeled neurons in the left hemicord corresponding to the ipsilateral (control) sciatic nerve. The higher the index, the greater was the axonal transport efficiency reduction.

Examination of the nerve stumps showed that HRP was taken up equally well by *both* nerve ends (Fig. 1). The tips of the transected nerves had the highest activity and this petered out in a disto-proximal fashion in the adjoining segments of nerves. Epineural and vascular structures were all intensely stained. In the spinal cords all animals revealed clear-cut differences in the staining of the control versus experimental side. Occasionally, spinal cord sections were totally free of labeled neurons in the experimental side (Fig. 1). Therefore, a complete and thorough search of *all* sections of a given specimen was necessary and revealed a variable number of labeled neurons in both sides of the cord. It was obvious, however, that labeled neurons in the left control half of any given specimen of spinal cord were more numerous and, on the whole, more intensely stained than labeled neurons in the right disulfiram-treated half of the *same* sample. This was confirmed by computing the afore-mentioned index in a series of five unselected consecutive rats: it varied between 0.50 and 0.80 and averaged 0.58 ± 0.12.

These findings indicate that epineurally applied disulfiram reduces retrograde axonal transport to about half of its original amount. Considering the possibility that impairment of axonal transport may be related to a clinically manifest neuropathy, the observed decrease in axonal transport may be regarded as an epiphenomenon of a structurally damaged nerve. On the other hand, clinical signs and structural changes of disulfiram develop on or after the third experimental day [13] and a limited electron microscopic study of test and control nerves reveals an indistinguishable ultra-structure over shorter periods of time when transport is already failing but changes

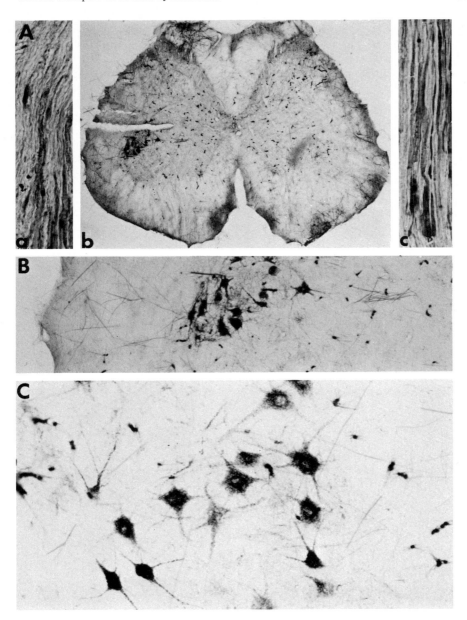

Fig. 1. Representative light micrographs of HRP activity in disulfiram treated and control sciatic nerves and corresponding spinal cord cell bodies. **A** Details of sections of the proximal stumps (most distal side at the bottom) showing comparable amounts of reaction product in segments lying next to transected tips of disulfiam-poisoned *(c)* and control *(a)* nerve trunks. × 205. Section of lumbar cord *(b)* with stained neurons (slit side = left side = control side) in dorsolateral nucleus of left ventral horn. × 21. **B** Detailed view of the left hemicord as shown above. × 65. **C** Higher magnification of variably labeled neurons in the control side of a spinal cord section adjacent to the one depicted above. × 205

and signs of neuropathy are not yet full-blown. An alternative explanation is that axonal transport may be reduced as a functional counterpart of a neuropathic damage without any cause and effect relationship to it. The third possibility is that impaired transport in disulfiram-poisoned nerve may actually induce structural changes in the treated axons. Since disulfiram and especially its diethyldithiocarbamate metabolite [5, 6] have a thiolprive effect on many sulfhydryl enzymes [4, 9], it is conceivable that the drug may vitiate axonal transport by inhibiting an adenosine triphosphatase instrumental in the force generating process.

Summary

We report here on the reduction of axonal transport after topical disulfiram exposure in the intact animal prior to the development of manifest neuropathy. Horseradish peroxidase was applied to the transected end of the sciatic nerve which in advance had been exposed to disulfiram. Retrogradely transported HRP was traced histochemically up to corresponding cell bodies at the spinal cord level.

References

1. Anzil AP (1980) Selected aspects of experimental disulfiram neuromyopathy. In: Manzo L et al (eds) Advances in neurotoxicology. Pergamon Press, Oxford New York, pp 359–366
2. Bouldin TW, Hall CD, Krigman MR (1980) Pathology of disulfiram neuropathy. Neuropathol Appl Neurobiol 6:155–160
3. Brimijoin S, Dyck PJ (1978) Axonal transport and peripheral nerve disease in man. Dev Neurol 1:139–146
4. DuBois KP, Raymund AP, Hietbrink BE (1961) Inhibitory action of dithiocarbamates on enzymes of animal tissues. Toxicol Appl Pharmacol 3:236–255
5. Faiman MD, Dodd DE, Hanzlik RE (1978) Distribution of S^{35} disulfiram and metabolites in mice, and metabolism of S^{35} disulfiram in the dog. Res Commun Chem Pathol Pharmacol 21:543–567
6. Gessner T, Jakubowski M (1972) Diethyldithiocarbamic acid methyl ester: a metabolite of disulfiram. Biochem Pharmacol 21:219–230
7. Graham RC, Karnovsky MJ (1966) The early stages of absorption of injected horseradish peroxidase in the proximal tubules of mouse kidney: ultrastructural cytochemistry by a new technique. J Histochem Cytochem 14:291–302
8. Hanson M, Edström A (1978) Mitosis inhibitors and axonal transport. Int Rev Cytol Suppl 7:373–402
9. Kitson TM (1975) The effect of disulfiram on the aldehyde dehydrogenases of sheep liver. Biochem J 151:407–412
10. Mesulam M-M (1978) Tetramethyl benzidine for horseradish peroxidase neurohistochemistry: a non-carcinogenic blue reaction-product with superior sensitivity for visualizing neural afferents and efferents. J Histochem Cytochem 26:106–117
11. Moddel G, Bilbao JM, Payne D, Ashby P (1978) Disulfiram neuropathy. Arch Neurol 35: 658–660
12. Ochs S (1975) Axoplasmic transport – a basis for neural pathology. In: Dyck PJ, Thomas PK, Lambert EH (eds) Peripheral neuropathy, vol I. Saunders, Philadelphia, pp 213–230
13. Zuccarello M, Anzil AP (1979) A localized model of experimental neuropathy by topical application of disulfiram. Exp Neurol 64:699–703

Section 4   Neuronal Pathology and Axoplasmic Transport

Axonal Transport in Human Nerve Disease and in the Experimental Neuropathy Induced by p-Bromophenylacetylurea

STEPHEN BRIMIJOIN[1], PETER JAMES DYCK[1], JOHANNES JAKOBSEN[1,2], and EDWARD H. LAMBERT[1]

As amply demonstrated elsewhere in this volume, the processes of rapid, slow, and retrograde axonal transport are essential to the growth and maintenance of nerve cells. It is therefore understandable that investigators have been attempting for over a decade to determine whether defects of transport underly any of the natural or toxin-induced diseases to which the peripheral nervous system is subject. Most of this effort has focused on the so-called "distal axonopathies" in which the early degeneration of the extremities of long neurons suggests the breakdown of an extended supply system [8]. Such neuropathies occur spontaneously in man [13] and can be induced by vita-min E deficiency [17] or by exposure to various toxic agents, e.g., acrylamide mono-mer [18], methyl-n-butyl ketone [23, 25], n-hexane [24] and organophosphate anti-cholinesterases [11].

Studies in experimentally induced distal neuropathies have not yet resolved the issue of the role of transport abnormalities. An early report by Pleasure et al. [21] indicated that transport was markedly impaired in the central processes of sensory neurons in cats exposed to acrylamide, but this was not confirmed by Bradley and Williams [2]. Neither Pleasure et al. nor Bradley and Williams observed significant effects of TOCP on transport, while James and Austin [16] failed to detect transport abnormalities in the neuropathy caused by the related agent, DFP. Mendell et al. [19] observed consistent slowing of fast transport in methyl-n-butyl ketone neuropathy, but the effect was small in relation to the associated pathology. The studies of Griffin, Price, and their collaborators on β,β'-iminodipropionitrile ([14]; see also Griffin et al., this volume) stand out in demonstrating a convincing link between impaired transport of neurofilament protein and the development of a neuropathy with filamentous axonal swellings. However this neuropathy involves primarily the proximal portion of axons and is not of the "dying-back" category.

Our own studies on transport in distal axonopathies began with biopsy samples of human sural nerve. Since rapid transport is a local process that can be sustained by isolated pieces of nerve, we utilized the in vitro redistribution of enzyme activity in ligated nerve samples as an index of the transport of unlabeled, endogenous protein (see [4] for a review of the uses and limitations of this methodology). Normative data were collected to establish the average content and velocity of transport of dopamine-

---

1   Departments of Pharmacology and Neurology, Mayo Clinic, Rochester, MN 55905, USA
2   Present adress: Second Clinic of Internal Medicine, University of Aarhus, Aarhus, Denmark

Axoplasmic Transport in Physiology and Pathology
(ed. by D.G. Weiss and A. Gorio)
© Springer-Verlag Berlin Heidelberg 1982

β-hydroxylase (DBH) and acetylcholinesterase (AChE) in the sural nerves of normal volunteer donors [5, 6]. In comparison with these control values, we noted deficits of enzyme content and accumulation in several types of hereditary and acquired neuropathy. Abnormalities were most marked in the conditions known to involve sensory and autonomic fibers, such as hereditary motor and sensory neuropathies (HMSN) types II and III, and diabetic neuropathy (Table 1, see also [6]).

Table 1. Enzyme content and transport in human sural nerve biopsy samples

Group	n	Dopamine-β-hydroxylase		Acetylcholinesterase	
		Content (units/mg wet weight)	Average velocity (mm/h)	Content (units/mg wet weight)	Average velocity (mm/h)
Controls	20	104 ± 15	1.8 ± 0.1	27 ± 4	1.0 ± 0.1
HSN I	4	64 ± 26	1.1 ± 0.4 [a]	–	–
HMSN II	4	74 ± 18	0.7 ± 0.5 [b]	–	–
HMSN III	9	106 ± 3	0.2 ± 0.1 [c]	–	–
Diabetic neuropathy	10	56 ± 11 [a]	0.6 ± 0.1 [c]	7.5 ± 1.7 [c]	0.5 ± 0.1 [b]

HSN = hereditary sensory neuropathy. HMSN = hereditary motor and sensory neuropathy (for nomenclature see [13]). Enzyme units are pmol octopamine produced per h (DBH) and nmol acetylcholine hydrolyzed per h (AChE).
[a] $P < 0.05$ vs control; [b] $P < 0.01$ vs control; [c] $P < 0.001$ vs control

Loss of nerve fibers is suspected to occur in most of these conditions [13]. Nevertheless, the reductions in enzyme accumulation have tended to be more severe than the changes in basal enzyme activity. This is particularly true of HMSN type III (Dejerine Sottas disease) in which the average DBH activity, in units per mg wet weight of nerve, has been found to be identical to the control mean, while the accumulation of DBH activity at a distal ligature was virtually nil (Table 1). A greater drop in accumulation than in basal content of enzyme activity results in a lower calculated average velocity of transport. Such a finding cannot be explained by fiber loss alone.

Diabetic neuropathy has particular interest since it is widespread and displays dying-back features (although proximal and "patchy" degeneration have been observed). Our largest series of neuropathic biopsy samples comes from patients with this disease. In these samples, the content and average transport velocity of both DBH and AChE were definitely abnormal (Table 1; [6]). To test whether the reduced average transport velocity reflected real slowing of transport we have carried out stop-flow experiments. These experiments depend on the general principle that rapid axonal transport can be locally interrupted in a reversible manner by locally exposing nerves to temperatures of 10°C or below. Upon rewarming of the cooled region, material that accumulated as a result of the transport block can be followed as a wave of increased concentration, which resumes migration along the axons [3]. The advantage of this approach is in permitting *direct* estimations of transport velocity in preparations like human nerve, in which the introduction of a radioactive label is not feasible.

Stop-flow techniques were applied to two normal and three diabetic human sural nerve biopsy samples in vitro. A short region of nerve (~ 5 mm) was cooled to approximately 5°C while the remainder of each sample was kept at 37°C in a simple three-compartment plexiglas chamber [3]. After 1.5 h of cooling, the samples were rewarmed to a uniform 37°C and transport was allowed to continue for another 1.5 h. At the end of this incubation, the samples were cut into consecutive 3-mm segments, which were assayed for DBH activity as previously described [6]. Normal nerves showed a peak of enzyme activity that had migrated 18 mm distally from the center of the cooled region (Fig. 1), corresponding to a transport velocity of 12 mm/h. This result is identical to one previously obtained in similar experiments on the normal rabbit sciatic nerve [3, 7]. A second peak of DBH activity remained at the center of the cooled region, presumably reflecting a degree of local damage induced by cooling. Diabetic nerve samples showed a much reduced distal peak of DBH activity, centering on the segment 9 mm from the cooled region (Fig. 1). This result means that the large reduction in "average transport velocity", calculated from the enzyme-accumulation in ligated nerve, probably does reflect an actual slowing of the transport of DBH along the axons of diabetic nerves.

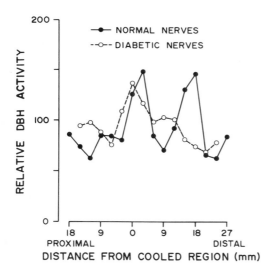

Fig. 1. Stop-flow experiment on DBH transport in normal (n = 2) and diabetic (n = 3) human sural nerve biopsy samples in vitro. Local cooling and rewarming periods each lasted 1.5 h

The great difficulty in interpreting the results of studies on transport in abnormal human nerve is to distinguish between primary and secondary events. Most of our biopsy samples have come from patients with advanced disease and might be expected to have ultrastructural alterations that could account for abnormalities in transport. Still, we have observed reduced transport in some conditions (e.g., early in uremia) in which axonal structure was probably normal. These findings have encouraged us to pursue studies of axonal transport in the early phases of the development of peripheral neuropathy. An experimental neuropathy that can be induced with consistent severity and timing by a chemical agent is obviously the most suitable object of this type of study.

One promising experimental neuropathy is caused in rats by *p*-bromophenylacetyl-urea (BPAU), a synthetic compound with no present industrial uses. BPAU neuropathy was discovered by Diezel and Quadbeck [12]. As elegantly described by Cavanagh et al. [9], the neuropathy is characterized by a curious latent period of up to 7 days after the administration of a single dose. During this period there are no clinical signs of neurologic deficit. Hindlimb weakness then develops rapidly. If the dose of BPAU is large enough, total hindlimb paralysis appears within an additional week. The fore-limbs are usually not involved, but hindlimb weakness can persist for 6 months or more.

Light microscopic studies have established that BPAU neuropathy belongs to the class of distal axonopathies. That is, the distal portions of long axons are the primary sites of pathological change [9]. Electron microscopy of affected regions has demon-strated that the distal axons are filled with "tubulomembranous" debris [1, 20]. This material resembles the accumulations of smooth endoplasmic reticulum that arise when rapid axonal transport is locally interrupted, as at a ligature or cold-block [26].

Since the smooth endoplasmic reticulum probably contains the majority of the proteins that move by fast axonal transport, the characteristic pathology of BPAU neuropathy would be consistent with a focal disturbance of rapid transport in distal axons. BPAU neuropathy thus presents an attractive opportunity for investigating the pathological significance of disturbances in transport.

Accordingly, we undertook a systematic examination of the effects of BPAU on axonal transport in the rat. This compound, which has not been commercially avail-able, was synthesized locally by Dr. Gerald Carlson of the Mayo Clinic Gastroenterol-ogy Unit. Product identity was confirmed by melting point determinations, and by infrared, and nuclear magnetic resonance spectroscopy, in addition to elemental analysis.

Single doses of BPAU, 200 mg/kg, were administered in dimethylsulfoxide (DMSO) solution to male Sprague-Dawley rats (300–400 g). Control rats received equivalent volumes of DMSO alone. Neurologic disability was evaluated by a modification of the clinical scoring system of Cavanagh et al. [9], dividing the rats arbitrarily into recognizable groups ranging from 0+ (no signs of disability) to 4+ (total hindlimb paralysis with involvement of forelimbs as well). By 15 days most rats had severe hindlimb weakness or total hindlimb paralysis (mean disability score, 2.7).

Measurements of rapid axonal transport were carried out at 2 and 15 days after administration of BPAU or DMSO. Injections of ^{35}S-methionine were made into the L5 dorsal root ganglia. The outflow of labeled protein was examined in unligated sciatic nerve, in order to assess transport velocity. Accumulation of radioactivity above and below a set of ligatures on the nerve was used as an index of the proximal and distal flux of labeled protein.

The maximal velocity of rapid axonal transport was entirely normal, even at 15 days after BPAU, when hindlimb paralysis had ensued [15]. Furthermore, BPAU neuropathy was not associated with any reduction in the anterograde flux of protein by rapid transport. In treated rats and controls, essentially the same amount of radio-activity accumulated on the proximal side of a set of ligatures placed on the sciatic nerve 8 h after injection of isotope. However, on the distal side of these ligatures, there was a large reduction in the accumulation of label in nerves of rats treated with

BPAU [15]. The reduction was proportionately greatest during the initial interval after injection (i.e., between 8 and 14 h after isotope). This result suggests a delay in the turnaround of proteins from rapid anterograde to rapid retrograde transport upon arrival in the distal axon or nerve terminal. However the reduction in accumulation was still present after collection intervals of 8 to 30 h, indicating that there was also a net decrease in the amount of protein ultimately recirculated in axons by retrograde transport.

It is especially noteworthy that the distal accumulation of ^{35}S-methionine labeled protein was already reduced within 2 days after BPAU administration [15]. This was early in the latent period, when no neurologic dysfunction could be detected "clinically" or even by electromyography (Lambert, Jakobsen and Brimijoin, unpublished). Because the turnaround abnormality preceded disability, it is worth considering the possibility that the abnormality is a causal link in the development of BPAU neuropathy.

Evidence favoring this possibility was obtained by comparing the severity of neuropathy with the degree of reduction of distal accumulation of labeled protein in each affected animal. As previously reported [15] there was a significant negative correlation between the clinical disability score and the amount of retrograde accumulation in a series of rats treated with BPAU, 200 mg/kg, 15 days previously ($r = -0.77$, $P < 0.05$). Since clinical scoring is notoriously prone to subjective error, we attempted to devise a more objective index of neurologic disability. For this purpose, rats were anesthetized with ether and subcutaneous bare needle recording electrodes were used to determine the maximal amplitude of the compound action potential of the peroneal muscles during supramaximal stimulation of the sciatic nerve. A good correlation was found between this measure and the independently assigned clinical disability score ($r = -0.82$; Jakobsen, Lambert, Carlson and Brimijoin, unpublished). Furthermore, as Fig. 2 shows, an even closer correlation was obtained between the retrograde accumulation and the amplitude of the muscle action potentials in BPAU treated rats. To our knowledge, these sorts of correlations have not previously been reported in studies on the role of axonal transport abnormalities in peripheral nerve disease.

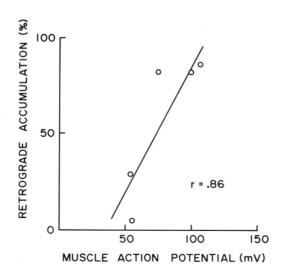

Fig. 2. Relation between severity of neurologic disability and reduction of retrograde transport in BPAU neuropathy. Rats were each given BPAU, 200 mg/kg, in DMSO solution, 15 days earlier. On the *horizontal axis* is the maximal amplitude of the compound action potentials in the peroneal musculature of the hindlimbs, recorded during supramaximal stimulation of the sciatic nerve, in situ. "Retrograde accumulation" is the percentage increase in amount of labeled protein distal to a set of ligatures in place during the period from 8–14 h after injection of the L5 dorsal root ganglion with ^{35}S-methionine. The correlation coefficient is statistically significant ($P < 0.05$)

On the basis of our results it is reasonable to entertain the hypothesis that the key lesion in BPAU neuropathy is an impairment of the recirculation of membrane-limited organelles delivered to the distal axon and terminals by rapid anterograde transport. We propose that this impairment leads directly to the distal accumulations of tubulo-membranous material reported by Blakemore and Cavanagh [1] and by Ohnishi and Ikeda [20]. Such accumulations could be expected to block the pathways for resupply, renewal, or loading of synaptic vesicles for release. The distorted geometry of the nerve terminal would thus impair synaptic transmission, causing muscular weakness and paralysis, as well as other functional deficits. If prolonged or extreme, distal accumulations of membranous debris could actually prevent the renewal of the terminal axolemma, which would eventually break down. In this way focal distal retention of transported materials could give rise to a Wallerian degeneration proceeding from the terminals and successively involving more proximal parts of the neuron, i.e., a "dying-back" neuropathy.

Our hypothesis is consistent with current findings in other experimental neuropathies. Mendell and his collaborators have demonstrated turnaround defects in a number of conditions, most notably zinc pyridinethione neuropathy [22]. These defects are associated with pathological changes that share features with BPAU neuropathy, i.e., intra-axonal accumulations of membranous debris. Even in acrylamide neuropathy, which has usually been considered to involve primarily accumulations of filamentous material [18], Droz's group have recently demonstrated a focal distal retention of rapidly transported protein that coincides with accumulated membranes of the smooth endoplasmic reticulum ([10] and Droz et al., this volume). We suspect that these focal accumulations are the morphological counterpart of impaired distal turnaround.

Thus, turnaround-defects, which represent disorders of the handling and disposition of rapidly transported materials in the distal axon, may underlie many distal axonopathies. These defects cannot be considered to represent derangements of axonal transport per se. For this reason, the role of transport abnormalities in distal axonopathy remains as problematical as ever. Nonetheless, turnaround is closely linked to transport and requires similar methodologies for its study. Further investigation of the turnaround abnormalities in BPAU neuropathy and other experimental conditions promises to shed light on the mechanisms of peripheral nerve disease in general.

Acknowledgments. We thank Dr. Gerald Carlson for his synthesis of BPAU. The expert technical assistance of Patricia Schreiber is also gratefully acknowledged. This work was supported by a Peripheral Nerve Disease Center Grant from the NIH (NS 14304) and by the Mayo Foundation.

References

1. Blakemore WF, Cavanagh JB (1969) Neuroaxonal dystrophy occurring in an experimental dying back process in the rat. Brain 92:789–804
2. Bradley WG, Williams MH (1973) Axoplasmic flow in axonal neuropathies. I. Axoplasmic flow in cats with toxic neuropathies. Brain 96:235–246

3. Brimijoin S (1975) Stop-flow: A new technique for measuring axonal transport, and its application to the transport of dopamine-β-hydroxylase. J Neurobiol 6:379–394

4. Brimijoin S (1982) Axonal transport in autonomic neurons: views on its kinetics. In: Kalsner S (ed) Trends in autonomic pharmacology, vol II. Urban & Schwarzenberg, München Baltimore, pp 17–42

5. Brimijoin S, Capek P, Dyck PJ (1973) Axonal transport of dopamine-β-hydroxylase by human sural nerves in vitro. Science 180:1295–1297

6. Brimijoin S, Dyck PJ (1979) Axonal transport of dopamine-β-hydroxylase and acetylcholinesterase in human peripheral neuropathy. Exp Neurol 66:467–478

7. Brimijoin S, Wiermaa MJ (1977) Direct comparison of the rapid axonal transport of norepinephrine and dopamine-β-hydroxylase activity. J Neurobiol 8:239–250

8. Cavanagh JB (1964) The significance of the "dying back" process in experimental and human nerve disease. Int Rev Exp Pathol 3:219–267

9. Cavanagh JB, Chen FCK, Kyu MH, Ridley A (1968) The experimental neuropathy in rats caused by p-bromophenylacetylurea. J Neurol Neurosurg Psychiatr 31:471–478

10. Chrétien M, Patey G, Souyri F, Droz B (1981) "Acrylamide-induced" neuropathy and impairment of axonal transport of proteins. II. Abnormal accumulations of smooth endoplasmic reticulum at sites of focal retention of fast transported proteins. Electron microscopic radioautographic study. Brain Res 205:15–28

11. Davies DR (1963) Neurotoxicity of organophosphorus compounds. In: Koelle GB (ed) Handbuch der experimentellen Pharmakologie, vol 15: Cholinesterase and anticholinesterase agents. Springer, Berlin Göttingen Heidelberg, pp 860–882

12. Diezel PB, Quadbeck G (1960) Nervenschädigung durch p-Bromophenylacetyl-Harnstoff. Naunyn-Schmiedebergs Arch Pharmakol Exp Pathol 238:534–541

13. Dyck PJ (1975) Inherited neuronal degeneration and atrophy affecting peripheral motor, sensory, and autonomic neurons. In: Dyck PJ, Thomas PK, Lambert EH (eds) Peripheral neuropathy. Saunders, Philadelphia. pp 825–867

14. Griffin JW, Price DL (1980) Proximal axonopathies induced by toxic chemicals. In: Spencer PS, Schaumburg HH (eds) Experimental and clinical neurotoxicology. Williams & Wilkins, Baltimore, pp 161–178

15. Jakobsen J, Brimijoin S (1981) Axonal transport of enzymes and labeled proteins in experimental axonopathy induced by p-bromophenylacetylurea. Brain Res 229:103–122

16. James KAC, Austin L (1970) The effect of DFP on axonal transport of protein in chicken sciatic nerve. Brain Res 13:192–194

17. Lampert P, Blumberg JM, Pentschew W (1964) An electron microscopic study of dystrophic axons in the gracile and cuneate nuclei of vitamin E deficient rats. J Neuropathol Exp Neurol 23:60–77

18. Le Quesne P (1980) Acrylamide. In: Spencer PS, Schaumburg HH (eds) Experimental and clinical neurotoxicology. Williams & Wilkins, Baltimore, pp 309–325

19. Mendell JR, Sahenk Z, Said F, Weiss HS, Savage R, Couri D (1977) Alterations of fast axoplasmic transport in experimental methyl-n-butyl ketone neuropathy. Brain Res 133:107–118

20. Ohnishi A, Ikeda M (1980) Morphometric evaluation of primary sensory neurons in experimental p-bromophenylacetylurea intoxication. Acta Neuropathol 52:111–118

21. Pleasure DE, Mishler KC, Engel WK (1969) Axonal transport of proteins in experimental neuropathies. Science 166:524–525

22. Sahenk Z, Mendell JR (1980) Axoplasmic transport in zinc pyridinethione neuropathy: Evidence for an abnormality in distal turn-around. Brain Res 186:343–353

23. Saida K, Mendell JR, Weiss HS (1976) Peripheral nerve changes induced by methyl-n-butyl ketone and potentiation by methyl ethyl ketone. J Neuropathol Exp Neurol 35:207–225

24. Schaumburg HH, Spencer PS (1976) Degeneration in central and peripheral nervous systems produced by pure n-hexane: an experimental study. Brain 99:183–192

25. Spencer PS, Schaumburg HH, Raleigh RL, Terhaar CJ (1975) Nervous system degeneration produced by the industrial solvent methyl-n-butyl ketone. Arch Neurol (Chicago) 32:219–222

26. Tsukita S, Ishikawa H (1980) The movement of membranous organelles in axons. Electron microscope identification of anterogradely and retrogradely transported organelles. J Cell Biol 84:513–530

Disturbances of Axoplasmic Transport in Alzheimer's Disease

PIERRE DUSTIN and JACQUELINE FLAMENT-DURAND [1]

In Alzheimer's disease, and in a small group of other neurological disorders, the normal microtubules and neurofilaments are replaced or displaced by an accumulation of so-called "paired helical filaments" (PHF) which are parallel oriented structures, periodically twisted every 80 nm. In light microscopy, they form the so-called "tangles" in the neuronal perikarya which stain with silver methods and show a strong birefringence with a green dichroism after Congo Red staining. This property is similar to that of amyloid and has led some authors to the false conclusion that these tangles represent intracellular deposits of amyloid.

In brains of patients affected with Alzheimer's disease, as in senility, in addition to the PHF, amyloid is found extracellularly, in the center of neuritic plaques. While the origin and mode of production of the PHF remain poorly understood, ultrastructural observation of nervous tissue presenting these changes clearly indicates that they are associated with disturbances of transport of various organelles. This may relate to the severe mental defects clinically observed, and deserve some comment in a book devoted to axoplasmic transport.

Material and Methods

By electron microscopy we have studied six brain biopsies from the frontal lobes of patients with Alzheimer's disease, and six fragments from brains of aged people, obtained post-mortem less than 6 h after death. The tangles are found in greatest numbers in the *cornu ammonis*. After paraffin embedding, they stain strongly with Congo Red, with the typical green birefringence, indicating that they are proteins in a β-pleated sheet configuration [5].

1 Department of Pathology and Electron Microscopy, Université Libre de Bruxelles, Brussels, Belgium

Axoplasmic Transport in Physiology and Pathology
(ed. by D.G. Weiss and A. Gorio)
© Springer-Verlag Berlin Heidelberg 1982

Results

The PHF are found in the neuronal perikarya and processes, axons and dendrites. They are often grouped in bundles which displace other neuronal structures. They are more frequent in non-myelinated processes (Fig. 1) but are also present in some myelinated fibers. Some of the processes containing PHF are almost entirely devoid of normal microtubules. However, normal synaptic junctions make contact with some of these fibers.

Evidence of disturbances of axoplasmic flow is provided by the accumulation of dense particles and swelling of the neuronal processes (Fig. 1). These aspects closely resemble those found above an experimental nerve ligation. The particles are of varied nature: many appear as altered mitochondria and some may be lysosomes.

Evidence that slow components of axoplasmic flow are affected is provided by the accumulation of dense particles, and in a few axons by important accumulations of apparently normal microtubules (Fig. 2).

Discussion

The characteristic lesion of Alzheimer's disease and the related form associated with senility (senile dementia of the Alzheimer type) is the formation of PHF in neurons. There may also be numerous senile plaques with amyloid deposits, but these are also found in aged people who are not demented, and in conditions not associated with old age (Down's syndrome), amyotrophic lateral sclerosis, kuru, and young drug addicts [3]. While PHF associated with senile plaques are mainly found in Alzheimer's disease and old age, they have also been described in some other conditions: the Guam-Parkinson dementia complex, Down's syndrome, dementia pugilistica, Haller-vorden-Spatz disease and subacute sclerosing panencephalitis. This last condition is most interesting, as it is a consequence of an atypical infection by measles virus [2, 13]. As recently pointed out by Wisniewski and Iqbal [17], a common feature links the core of senile plaques and PHF: they stain by Congo Red with a green dichroism in polarized light, indicating that both belong to the group of β-pleated sheet proteins [12].

It is interesting to mention in relation with this fact, that the bundles of intermediate (10 nm) filaments induced in several types of cells after colchicine treatment have been shown to stain with Congo Red like amyloid, hence they may have also a β-pleated sheet structure [11].

The biochemistry of PHF remains poorly understood. Although described since 1963 by Terry [15] — who already pointed out their resemblance to amyloid — they are very rarely found in animals and many attempts to reproduce similar structures have been unsuccessful. In old Rhesus monkeys with senile plaques Wisniewski et al. [18] found some similar structures in dendrites, although the periodicity of the twisting was 50 nm, differing from that in man (80 nm). In rats aged 11 months,

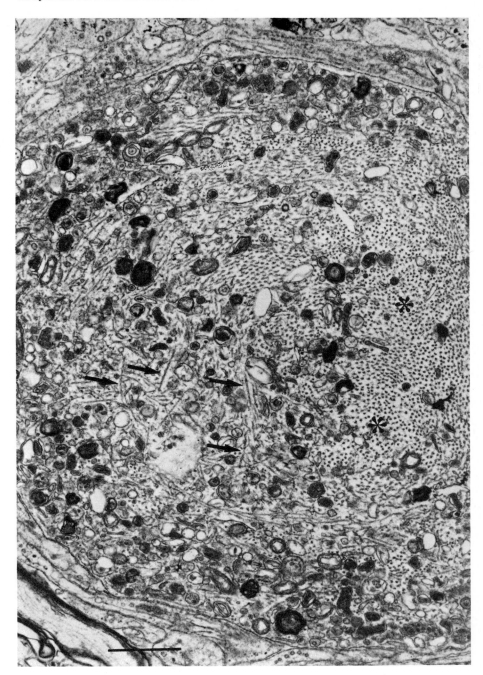

Fig. 1. Brain biopsy in a case of Alzheimer's disease, aged 65 years. Unmyelinated process considerably swollen with large numbers of paired helical filaments (PHF) cut longitudinally ↓ and transversally *. Numerous lamellar dark bodies are visible (altered mitochondria or lysosomes). Bar 1 μm

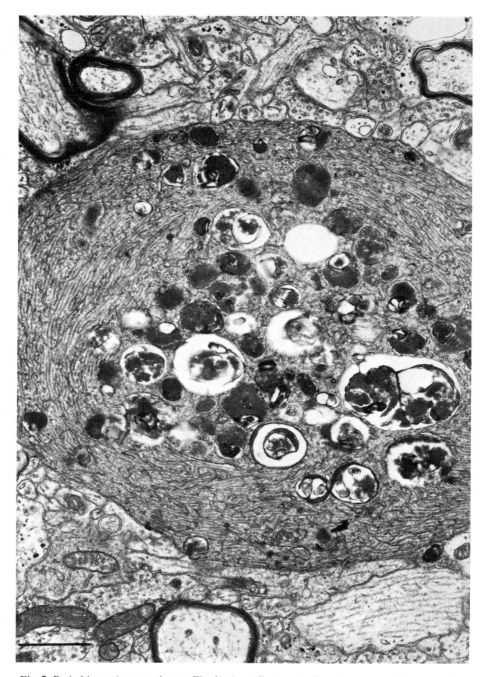

Fig. 2. Brain biopsy (same patient as Fig. 1). A swollen, unmyelinated process, with a central accumulation of dense bodies and at the periphery a large number of morphologically normal microtubules. These are abnormally numerous, and their circular disposition is pathological. Close to this abnormal process, several normal, myelinated axons are visible. Bar 1 μm

who drank water with 15% ethyl alcohol for five months, spinal neurones showed structures similar to the PHF [16], but the author did not mention whether they stained with Congo Red. De Boni and Crapper [1] observed a few pairs of intermediate filaments (diameter: 9.5 nm) irregularly twisted two by two with crossings spaced from 80–160 nm, in cultured neurons of human fetuses, under the influence of a saline extract of cortex of Alzheimer patients. However, some were also found in controls.

An interesting finding is that described in the brains of aged rats: in one animal, aged 27 months, of the Wistar-Kyoto strain, typical PHF were observed. They formed bundles in axons and differed from those of human brains only in their shorter period-icity (34 nm) [10].

Biochemical studies of the proteins of brains with PHF remain controversial, and it is not known whether they originate from microtubules, neurofilaments or other proteins. A 50,000 dalton polypeptide was isolated by Iqbal et al. [6]. This may originate from neurotubules, as an antiserum to microtubules stained the tangles by immunofluorescence and immunohistochemistry [6–8]. On the contrary, Ishii et al. [9] obtained results by immunofluorescence which indicated a neurofibrillary origin of the tangles of Alzheimer's disease. In a discussion of these data, Wisniewski and Iqbal [17] conclude that it is possible that no new proteins appear in the tangles, but that some factors induce the β-pleated sheet configuration of neurofilament or microtubule proteins. On the contrary, a biochemical study of isolated cortical neurones from Alzheimer's disease, with abundant PHF, showed an increase of a 20,000 dalton protein, while tubulins and neurofilament proteins were normal [14]. The origin of this protein remains unclear. The technique of isolation of affected neurons used by these authors is certainly promising.

It is evident that morphological data, such as those presented above, cannot alone solve the problem of the origin of the PHF. They do, however, indicate a marked decrease in the numbers of normal microtubules and neurofilaments in the affected cells. Moreover, this is linked with evidence of severe disturbances of axonal flow as indicated by the accumulation of osmiophilic bodies and sometimes of microtubules. These findings suggest that most forms of axonal flow are inhibited, including the slow form by which microtubules are known to be transported. The impaired flow may perhaps explain the main metabolic changes which are known so far in Alzhei-mer's disease, i.e. the decrease of choline acetyl transferase and of acetylcholine esterase [4]. The morphological evidence of arrested neuroplasmic flow should be considered in studies on the relation between the neuronal changes and the clinical symptoms of Alzheimer's disease and senile dementia.

Summary

In brains of patients affected with Alzheimer's disease and in brains of some aged patients the cortical neurons of the brain show accumulation of paired helical fila-ments (PHF). This change is related to evidence of severe disturbances of axoplasmic flow. The numbers of microtubules and neurofilaments appear strongly decreased in the affected cells. The conflicting theories about the origin of the PHF are discussed.

Acknowledgments. Part of this work was made possible with the help of credit no. 3.4512.80 of the Belgian National fund for Medical Research.

References

1. De Boni U, Crapper DR (1978) Paired helical filaments of the Alzheimer type in cultured neurones. Nature (London) 271:566–568
2. Dubois-Dalcq M, Worthinton K, Gutenson O, Barbosa LH (1975) Immunoperoxidase labelling of subacute sclerosing panencephalitis virus in hamster acute encephalitis. Lab Invest 32:518–526
3. Dustin P, Flament-Durand J, Vandermot G (1979) Aspects ultrastructuraux du viellissement cérébral. Bull Mém Acad R Med Belg 134:369–383
4. Dziedzic JD, Iqbal K, Wisniewski HM (1980) Central cholinergic activity in Alzheimer dementia. J Neuropathol Exp Neurol 39:351
5. Glenner GG (1980) Amyloid deposits and amyloidosis. The β-fibrilloses. N Engl J Med 302: 1283–1292, 1333–1343
6. Iqbal K, Grundke-Iqbal I, Wisniewski HM, Terry RD (1978) Neurofibers in Alzheimer dementia and other conditions. In: Katzman R, Terry RD, Bick KL (eds) Alzheimer's disease, senile dementia and related disorders. Raven Press, New York, pp 409–420
7. Iqbal K, Grundke-Iqbal I, Johnson AB, Wisniewski HM (1979) Are neurotubule proteins involved in the formation of Alzheimer neurofibrillary tangles? J Neuropathol Exp Neurol 38:322
8. Iqbal K, Grundke-Iqbal I, Johnson AB, Wisniewski HM (1980) Neurofibrous proteins in aging and dementia. In: Amaducci L, Davison AN, Antuono P (eds) Aging of the brain and dementia. Raven Press, New York, pp 39–48
9. Ishii T, Haga S, Tokutake S (1979) Presence of neurofilament protein in Alzheimer's neurofibrillary tangles (ANT). An immunofluorescent study. Acta Neuropathol 48:105–112
10. Knox CA, Yates RD, Chen IL (1980) Brain aging in normotensive and hypertensive strains of rats. II. Ultrastructural changes in neurons and glia. Acta Neuropathol 52:7–16
11. Linder E, Lehto V-P, Virtanen I (1979) Amyloid-like green birefringence in cytoskeletal 10 nm filaments after staining with Congo Red. Acta Pathol Microbiol Scand 87:299–306
12. Luse SA, Smith KR Jr (1964) The ultrastructure of senile plaques. Am J Pathol 44:553–563
13. Mandybur TI, Nagpaul AS, Pappas Z, Niklowitz WJ (1977) Alzheimer neurofilbrillary changes in subacute sclerosing panencephalitis. Ann Neurol 1:103–107
14. Selkoe DJ (1980) Altered protein composition of isolated human cortical neurons in Alzheimer disease. Ann Neurol 8:468–478
15. Terry RD (1963) The fine structure of the neurofibrillary tangles in Alzheimer's disease. J Neuropathol Exp Neurol 22:629–642
16. Volk B (1980) Paired helical filaments in rat spinal ganglia following chronic alcohol administration: an electron microscopic investigation. Neuropathol Appl Neurobiol 6:143–153
17. Wisniewski HM, Iqbal K (1980) Aging of the brain and dementia. Trends Neurosci 3:226–228
18. Wisniewski HM, Ghetti B, Terry RDJ (1973) Neuritic (senile) plaques and filamentous changes in aged rhesus monkeys. J Neuropathol Exp Neurol 32:566–584

Axonal Transport of Acetylcholinesterase Molecular Forms in Sciatic Nerve of Genetically Diabetic Mice

MAURIZIO VITADELLO [1,2], JEAN-YVES COURAUD [1], RAYMONDE HÄSSIG [1], ALFREDO GORIO [2], and LUIGI DI GIAMBERARDINO [1]

The mutant "diabetes" in the C57 BL/Ks inbred mouse (CNRS, Orleans, France) is an autosomal recessive with full penetrance in the homozygote [2, 7]. The homozygote mouse is characterized by metabolic disturbances resembling diabetes mellitus in man: at the age of 4 weeks the animal is hyperglycemic, hyperinsulinemic, hyperphagic and obese. At 4—5 months it becomes severely hyperglycemic, with hypoinsulinemia, weight loss and early death [6, 9].

Electrophysiological and morphometric studies on peripheral nerves [5, 8, 11] revealed that sensory and motor conduction velocities are decreased, this alteration being associated with reduction in size of large myelinated fibres in long standing diabetes. These alterations can be considered as the early symptoms of an axonal neuropathy.

Because maintenance of axons requires a continuous supply of macromolecules synthesized in neuronal perikarya, it was of interest to investigate whether alterations of axoplasmic transport were present in this animal model of diabetic neuropathy. To this purpose we studied the axonal transport of acetylcholinesterase (AChE), an enzyme which is known to be transported down peripheral nerves of avians [3] and mammals [4, 12], in diabetic mice of various ages (50—180 days).

Axonal transport of AChE was tested with the nerve-section method [3] by measuring the amount of enzyme accumulated over a 24 h period in the 2 mm nerve segment proximal to a test section (see legend to Fig. 2).

AChE transport in diabetic (db/db) mice was identical to that measured in control (db/−) heterozygote mice up to the age of 130 days, then a 20% reduction was noticed at 180 days: at this age the amount of AChE enzymatic activity at the test section was 0.94 ± 0.1 U/24 h in db/db and 1.12 ± 0.1 U/24 h in db/− mice. Sedimentation analysis revealed that the 20% drop was entirely due to a reduced accumulation of G_1 and G_2 molecular forms of AChE, while the accumulation of G_4 and A_{12} remained unchanged (Figs. 1 and 2). As G_4 and A_{12} are known to be transported by rapid axoplasmic flow, our observation suggests that in mouse diabetic neuropathy fast axoplasmic transport was not affected up to the age of 180 days. The reduced accumulation of light molecular forms G_1 and G_2, which are most likely conveyed [3] by the slow axoplasmic transport, could be explained either by a reduced velocity or by a diminished clearance of the slow axonal transport of these forms.

1   Département de Biologie, C.E.N. de Saclay, 91191 Gif-sur-Yvette Cedex, France
2   FIDIA Research Laboratories, 35031 Abano Terme, Italy

Axoplasmic Transport in Physiology and Pathology
(ed. by D.G. Weiss and A. Gorio)
© Springer-Verlag Berlin Heidelberg 1982

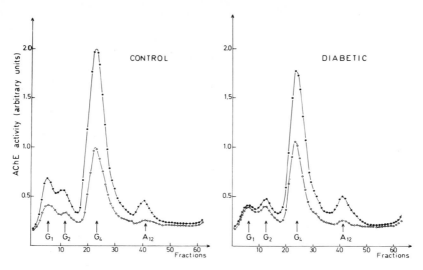

Fig. 1. Sedimentation profiles of AChE in intact sciatic nerve ($-\circ-$) and in the 2 mm nerve segment proximal to a test section ($-$. $-$) of db/db (diabetic) and db/$-$ (control) mouse at the age of 180 days. The sciatic nerve was cut with fine scissors at 5 mm after the ischiatic foramen and a 2 mm nerve segment was resected to measure the normal level of AChE. 24 h later the 2 mm segment at the tip of the proximal stump was dissected. Nerve segments from 4 animals were pooled and homogenized in 300 μl of extraction buffer (1 M NaCl, 0.01 M phosphate buffer pH 7.0, 5 mM EGTA, 1% Triton X100, 2 mg/ml bacitracin, 32 mg/ml benzamidin). The homogenates were centrifuged at 20,000 g for 20 min, and usually 20 μl of the supernatant were assayed for AChE activity in 1 ml of Ellman medium (see [3]). The remainder of the supernatant was layered on 5$-$20% sucrose gradient in extraction buffer and run at 250,000 g for 17 h, at 4°C in a SW-41 Beckman or a TST-41 Kontron rotor. All fractions were assayed for AChE activity in Ellman medium. Four molecular forms of AChE were separated. Their sedimentation coefficients were respectively 4S (G_1), 6S (G_2), 10S (G_4) and 16S (A_{12}). A reduced accumulation of G_1 and G_2 molecular forms can be noted in sectioned diabetic sciatic nerve compared to control

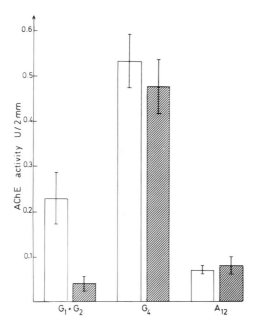

Fig. 2. Distribution of AChE molecular forms accumulated at the test-section (see Fig. 1) in db/db *(grey columns)* and db/$-$ mouse, at the age of 180 days. The values presented here were corrected for the concentration normally present in the nerve. Note the decrease in accumulated G_1 and G_2 AChE molecular forms in db/db transected sciatic nerve compared to control db/$-$ ($P \geqslant 0.01$), while the accumulation of G_4 and A_{12} was essentially unchanged. Data are average of 3 to 5 determinations ± S.E.M.; each determination was done on nerve segments pooled from 4 animals

Interestingly no alteration of the transport of AChE was detectable up to 130 days of age in diabetic mice, in spite of Sima and Robertson's report of a significant reduction of nerve conduction velocity [11]. On the contrary, the axonal dwindling of peripheral nerve previously described in 185 days old db/db mice [5] coincides with our observation of a reduced transport of the G_1 and G_2 AChE forms, suggesting that these two phenomena could be correlated. In fact it would not be surprising that a decrease of slow axonal transport could induce a reduction of axonal cross-section.

Acknowledgments. This work was done during a program, directed by L. Di Giamberardino cosponsored by the Commissariat à l'Energie Atomique (France) and the FIDIA S.p.A. (Abano Terme, Italy) and realised at the Département de Biologie, C.E.N. de Saclay, where Dr. Vitadello is staying as a visiting scientist since November 1980.

References

1. Brimijoin S (1979) Axonal transport and subcellular distribution of molecular forms of acetylcholinesterase in rabbit sciatic nerve. Mol Pharmacol 15:641–648
2. Coleman DL, Hummel KP (1967) Studies with the mutation diabetes in the mouse. Diabetologia 3:238–248
3. Couraud JY, Di Giamberardino L (1980) Axonal transport of the molecular forms of acetylcholinesterase in chick sciatic nerve. J Neurochem 35:1053–1066
4. Fernandez HL, Duell MJ, Festoff BN (1980) Bidirectional axonal transport of 16 S AChE in rat sciatic nerve. J Neurobiol 11:31–39
5. Gorio A, Aporti F, Norido F (1981) Ganglioside treatment in experimental diabetic neuropathy. In: Rapport A, Gorio A (eds) Gangliosides in neurological and neuromuscular function, development and repair. Raven Press, New York, pp 259–266
6. Herberg L, Coleman DL (1977) Laboratory animals exhibiting obesity and diabetes syndrome. Metabolism 26:59–99
7. Hummel KP, Dickie MM, Coleman DL (1966) Diabetes, a new mutation in the mouse. Science 153:1127–1128
8. Moore SA, Peterson RG, Felten DL, Cartwright TR, O'Connor BL (1980) Reduced sensory and motor conduction velocity in 25-week-old diabetic [C57 BL/Ks (db/db)] mice. Exp Neurol 70:548–555
9. Mordes JP, Rossini AA (1981) Animal models of diabetes. Am J Med 70:353–360
10. Schmidt RE, Matschinsky FM, Godfrey DA, Williams AD, McDougal BB Jr (1975) Fast and slow axoplasmic flow in sciatic nerve of diabetic rats. Diabetes 24:1081–1085
11. Sima AF, Robertson DM (1978) Peripheral neuropathy in mutant diabetic mouse [C57 BL/Ks (db/db)]. Acta Neuropathol 41:85–89
12. Skau KA, Brimijoin S (1980) Multiple molecular forms of acetylcholinesterase in rat vagus nerve, smooth muscle and heart. J Neurochem 35:1151–1154

Effects of Graded Compression
on Axonal Transport in Peripheral Nerves

JOHAN SJÖSTRAND[1], W. GRAHAM McLEAN[2], and BJÖRN RYDEVIK[3]

Acute and chronic compression of peripheral nerves may result in disorders of sensory and motor function. Clinically, this is seen in association with e.g., trauma to extremities and nerve entrapment syndromes. It has been shown experimentally that compression of a peripheral nerve may entail structural damage to myelin and axons [11] as well as impairment of intraneural microcirculation [16]. The reaction of the axonal transport systems in association with nerve compression lesions is, however, comparatively unknown. This chapter reviews some experimental studies in which the effects of graded compression on axonal transport have been studied. The experimental data will be discussed in relation to the pathophysiology of compression injury of peripheral nerves.

Methods

Compression

Several different models for experimental compression of peripheral nerves have previously been described, e.g. indirectly as whole limb compression by means of a tourniquet [2, 4, 8, 11], or directly by various kinds of pressure-devices compressing an exposed nerve trunk [1, 3]. We have used a device which was specially developed to allow compression of peripheral nerves at defined, graded pressure levels [13, 15]. By this method direct compression can be applied to exposed peripheral nerves by means of an inflatable "mini-cuff", comprising two symmetrical halves made of plexiglass. Two thin rubber membranes are glued on each plexiglass half and the nerve is compressed between the two membranes (Fig. 1). The compression device is connected to a compressed air system, allowing inflation at graded, controlled pressure

1   Institute of Neurobiology, Department of Ophthalmology, University of Göteborg, S-41345 Göteborg, Sweden
2   Department of Pharmacology and Therapeutics, University of Liverpool, Liverpool L69 3BX, England
3   Laboratory of Experimental Biology, Department of Anatomy and Department of Orthopaedic Surgery I, University of Göteborg, S-41345, Göteborg, Sweden

Axoplasmic Transport in Physiology and Pathology
(ed. by D.G. Weiss and A. Gorio)
© Springer-Verlag Berlin Heidelberg 1982

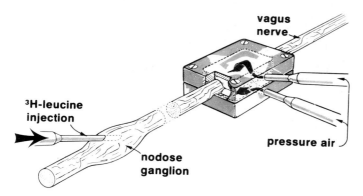

Fig. 1. Schematic drawing of experimental procedure. Axonally transported proteins were labelled by injection of ^{3}H-leucine into the nodose ganglion of the vagus nerve. In acute experiments the labelling was performed 2 h before application of the compression chamber around the nerve. The chamber, consisting of plexiglass and rubber membranes, was inflated with air to a known pressure (20–400 mm Hg) for 2 h. In other experiments, a recovery period from 1–14 days after the compression was allowed before labelling of proteins was performed. (From [15], reproduced by kind permission of the Editor of J. Neurol. Neurosurg., Psychiatry)

levels. The magnitude of the compression-trauma inflicted to the nerve is thus determined by the pressure level and the duration of the compression. The mini-cuff was applied around the rabbit cervical vagus nerve about 20–30 mm from the nodose ganglion. The effects on rapid axonal transport in this nerve of 20, 30, 50, 200 and 400 mm Hg applied for 2 h, have been tested [15, 17].

Axonal Transport

In the above mentioned studies, the rapid axonal transport was measured as follows: The nodose ganglion was exposed and 200 μl (100 μCi) of ^{3}H-leucine in 0.9% NaCl were injected into the nodose ganglion of the vagus nerve. In acute experiments, pressure was applied to the vagus nerve 2 h after labelling with ^{3}H-leucine; in others there was a recovery-period from 1–14 days before protein labelling. The animals were sacrificed 4 h after injection of the isotope. The nodose ganglion and the vagus nerve down to the thoracic cavity were immediately removed and placed on ice. The nerve was then cut into 2,5-mm long pieces and the trichoroacetic acid-precipitable radioactivity was measured as described [9]. The radioactivity of each nerve piece was plotted against the distance of the piece from the nodose ganglion. This gave a profile of the distribution of labelled proteins in the vagus nerve 4 h after injection of ^{3}H-leucine into the nodose ganglion. In untreated nerves, a wave-front of labelled proteins appeared about 60 mm from the ganglion, indicating a rate of axonal transport of at least 360 mm/day, in agreement with [9]. In nerves in which compression had blocked axonal transport, an accumulation of labelled proteins was found in the nerve in the region of the compression. By drawing profiles of the radioactivity in nerves from each experiment, the extent to which compression had blocked axonal transport was estimated (Fig. 2).

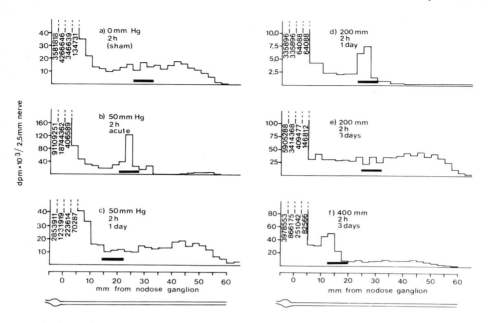

Fig. 2 a–f. Summary of typical findings in various experimental series. Pressure, time of application, and recovery period is shown in each diagram. A *black bar* indicates the site of compression. Application of the chamber around the nerve for 2 h without inflation (sham-experiments) caused no, or just a minimal, accumulation of axonally transported proteins (a), in contrast to 50 mm Hg applied for 2 h which caused a block of axonal transport (b). This block was reversible within one day (c). In the experiment shown in d, a block of axonal transport was persisting 1 day after compression at 200 mm Hg for 2 h. After a 3 day recovery period, normal transport was found (e). In cases of compression at 400 mm Hg for 2 h a blockage of transport was seen even 3 days after the trauma (f). (From [15], reproduced by kind permission of the Editor of J. Neurol. Neurosurg. Psychiatry)

Results and Comments

Axonal Transport During Compression

The studies reported here have demonstrated that the rapid anterograde axonal transport of the rabbit vagus nerve is acutely blocked by compression during 2 h at pressure levels from 400 mm Hg down to 30 mm Hg [15, 17] (Figs. 2, 3). However, 20 mm Hg or application of the chamber without inflation, for 2 h did not significantly affect axonal transport (Figs. 2, 3). There thus seems to be a lower critical pressure limit at 30 mm Hg which, when applied for 2 h, blocks the rapid axonal transport. The mechanism behind the blockage of axonal transport induced by compression in this manner is not fully known. Theoretically, compression may affect axonal transport in two ways: either by mechanical deformation of nerve fibers or by interference with intraneural blood flow, i.e. the oxygen supply to the nerve fibers

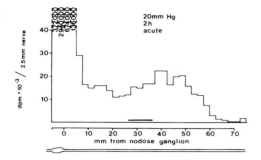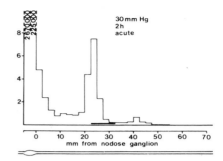

Fig. 3. Demonstration of the findings at the low pressure levels, 20 mm Hg and 30 mm Hg, applied for 2 h. Note that there is no significant effect on axonal transport at 20 mm Hg, but when the nerve is compressed at 30 mm Hg for 2 h there is a marked block of axonal transport

[7, 12, 15]. At the lower pressure levels in these experiments (30–50 mm Hg) there is a significant impairment of intraneural blood flow as seen in vital microscopic observations of the intraneural microcirculation during compression [16]. Some degree of nerve fiber deformation may however, also occur during compression at these pressure levels [6]. Compression at higher pressure levels (200–400 mm Hg) involves complete ischemia of the compressed nerve segment [16] but also severe nerve fiber deformation [14] and these factors both probably contribute to the pronounced axonal transport block in these cases.

Reversibility of Axonal Transport Blockage

Our experiments have demonstrated marked differences in the recovery time from the block of axonal transport after compression at 50, 200 and 400 mm Hg, applied for 2 h [15] (Fig. 2). In nerves which had been compressed at 50 mm Hg for 2 h axonal transport recovered to normal values within the first day after compression. This rapid reversibility of axonal transport block indicates that the cause of the accumulation induced at this pressure level is ischemia [15]. Following pressure release in experiments where 200 mm Hg and 400 mm Hg had been applied for 2 h there was a prolonged impairment of axonal transport persisting up to 1 and 3 days respectively after the trauma. The delayed recovery of axonal transport in these cases may be an effect of mechanical injury to the nerve fibers [14]. However, there is also evidence that compression of this magnitude may induce endoneurial edema [13], which in turn may lead to an increase of the endoneurial fluid pressure (EFP) [10]. Such increase of EFP might cause compression of the axons and the endoneurial blood vessels and consequently directly or indirectly affect axonal transport. A posttraumatic endoneurial edema may also alter the local electrolyte balance around the axons, and may by such a mechanism also contribute to the impairment of axonal transport.

Clinical Aspects

Recent investigations have demonstrated that in patients with carpal tunnel syndrome, the median nerve in the carpal tunnel is subjected to compression at a pressure level of about 30 mm Hg [5]. This pressure level corresponds nicely to the critical pressure at which we have found blockage of axonal transport. It thus seems justified to assume that axonal transport may be impaired in connection with nerve entrapment syndromes. The possible role of axonal transport blockage for the development of changes in axonal structure and function in these situations is, however, not known, but deserves further investigation.

Acknowledgments. The studies reviewed here have been supported by grants from the Swedish Medical Research Council (projects no. 5188 and 2226), the Swedish Work Environment Fund, the Göteborg Medical Society and the University of Göteborg.

References

1. Bentley FH, Schlapp W (1943) The effects of pressure on conduction in peripheral nerve. J Physiol (London) 102:72–82
2. Denny-Brown D, Brenner C (1943) Paralysis of nerve induced by direct pressure and by tourniquet. Arch Neurol Psychiatry 51:1–26
3. Denny-Brown D, Brenner C (1944) Lesion in peripheral nerve resulting from compression by spring clip. Arch Neurol Psychiatry 52:1–19
4. Fowler RJ, Danta G, Gilliatt RW (1972) Recovery of nerve conduction of a pneumatic tourniquet: Observations of the hind limb of the baboon. J Neurol Neurosurg Psychiatry 35: 638–647
5. Gelberman R, Hergenroeder P, Hargens AR, Lundborg G, Akeson W (1981) The carpal tunnel syndrome. A study of carpal tunnel pressures. J Bon J Surg 63A:380–383
6. Hahnenberger RW (1978) Effects of pressure on fast axoplasmic flow. Acta Physiol Scand 104:299–308
7. Leone J, Ochs S (1978) Anoxic block and recovery of axoplasmic transport and electrical excitability of nerve. J Neurobiol 9:229–245
8. Lundborg G (1970) Ischemic nerve injury. Experimental studies on intraneural microvascular pathophysiology and nerve function in a limb subjected to temporary circulatory arrest. Scand J Plast Reconstr Surg, Suppl 6
9. McLean WG, Frizell M, Sjöstrand J (1976) Slow axonal transport of proteins in sensory fibers of rabbit vagus nerve. J Neurochem 26:1213–1216
10. Myers RR, Powell HC, Shapiro HM, Costello ML, Lampert PW (1980) Changes in endoneurial fluid pressure, permeability, and peripheral nerve ultrastructure in experimental lead neuropathy. Ann Neurol 8:392–401
11. Ochoa J, Fowler TJ, Gilliatt RW (1972) Anatomical changes in peripheral nerves compressed by a pneumatic tourniquet. J Anat 113:433–455
12. Ochs S (1974) Energy metabolism and supply of ~ P to the fast axoplasmic transport mechanism in nerve. Fed Proc 33:1049–1058
13. Rydevik B, Lundborg G (1977) Permeability of intraneural microvessels and perineurium following acute, graded nerve compression. Scand J Plast Reconstr Surg 11:179–187
14. Rydevik B, Nordborg C (1980) Changes in nerve function and nerve fibre structure induced by acute, graded compression. J Neurol Neurosurg Psychiatry 43:1070–1082

15. Rydevik B, McLean WG, Sjöstrand J, Lundborg G (1980) Blockage of axonal transport induced by acute, graded compression of the rabbit vagus nerve. J Neurol Neurosurg Psychiatry 43:690–698
16. Rydevik B, Lundborg G, Bagge U (1981) Effects of graded compression on intraneural blood flow. J Hand Surg 6:3–12
17. Rydevik B, Dahlin L-B, Danielsen N, McLean WG, Sjöstrand J (1982) Impairment of rapid axonal transport during compression at low pressure. An in vivo study on the rabbit vagus nerve. (In preparation)

Role of Axonal Transport in the Pathophysiology of Retinal and Optic Nerve Disease

JOHAN SJÖSTRAND[1]

Introduction

The analysis of intra-axonal transport of retinal ganglion cells in various retinopathies and optic neuropathies has given a unique insight into the dynamics of the axonal transport and metabolism of the diseased retinal ganglion cells. In a number of recent studies axonal transport dysfunctions have been described in association with a variety of pathological conditions affecting the retinal ganglion cell and its axon.

In this short review I will present some data concerning axonal transport obstruction in retinal ganglion cells under pathological conditions in an attempt to discuss some recent ideas concerning the cellular pathophysiology in disorders affecting the retinal ganglion cell and its axon.

It is a well-known fact that complete blockage of transport, for example by separating the cell body from its axon by axotomy, will cause axon degeneration below the site of lesion. This axonal degeneration probably occurs following all kinds of prolonged complete blockage and is thought to be the consequence of blocked supply of macromolecules and organelles from the cell body. Experiments with transport-blocking drugs, such as colchicine injected into the eye, have demonstrated that both rapid and slow transport of the retinal ganglion cells are profoundly inhibited. The effect of this subtotal transport blockage on the morphological and/or electrophysiological properties have been investigated in the retrobulbar optic pathway (for review, see [5]). These findings imply that the transmission mechanism in the synaptic terminals in the long term are dependent on supply of material by axonal transport. The electrical impulse propagation along the axons, however, does not seem to be directly associated with the axonal transport. The experiments also showed that drug-induced, severe transport block may be reversible. In experiments using local, graded compression on a peripheral nerve it has been shown that the axons may survive even a complete blockage of axonal transport lasting at least a day. Following prolonged, complete transport blockades axon degeneration starts to occur [16]. These experiments raise the following important questions:

1 Department of Ophthalmology, University of Göteborg, Sahlgren's Hospital, S-41345 Göteborg, Sweden

Axoplasmic Transport in Physiology and Pathology
(ed. by D.G. Weiss and A. Gorio)
© Springer-Verlag Berlin Heidelberg 1982

1. Which types of transport dysfunction may play a role in the pathogenesis of various optic neuropathies?
2. How sensitive are the axons and their terminals to transport obstruction of the rapid and/or slow phases?
3. How chronic must these changes be to cause functional deterioration or to cause axonal degeneration?

To define the role of axonal transport dysfunction in the pathogenesis of optic neuropathies these questions have to be answered.

Ischemic Blockage of Axonal Transport

Axonal transport is dependent on the local supply of energy to the axon [11]. Occlusion of any of the retinal arterioles supplying the nerve fiber layer of the retinal ganglion cells may therefore block the axonal transport. This has been demonstrated in an experimental study of pig retina following occlusion of small retinal arterioles by argon laser photocoagulation [7]. On both sides of the zone with ischemic necrosis caused by the arteriole obstruction swollen axonal endings developed filled with organelles. These swellings gave rise to whitish, opaque areas. Autoradiography demonstrated accumulation of rapidly transported radiolabelled proteins exported from the retinal ganglion cell bodies to the regions of axonal swellings. These experiments indicate obstruction both of the anterograde and retrograde axonal transport (Fig. 1). The cotton-wool spots therefore develop from the accumulations in and

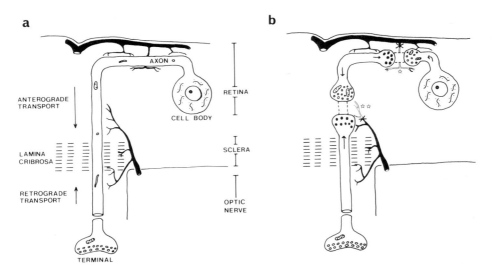

Fig. 1 a, b. Schematic drawing of the axonal transport pattern within the retinal ganglion cell under normal conditions (a) and following arteriolar obstruction (b) at the level of the retina (☆) or optic nerve head (☆☆). Above and below the ischemic zone macromolecules and organelles transported with both the anterograde or retrograde flow are accumulating

swelling of the axons central and peripheral to the ischemic area. The width of the cotton-wool spot area seems to depend on the distance that different axons penetrate into the ischemic area before the axonal transport is obstructed [7]. Similar axonal swellings with associated obstruction of rapid anterograde transport (Fig. 1) and subsequent pale swelling of the optic disc have been demonstrated in the optic nerve of the monkey following occlusion of the arteries supplying that area [8]. Clinical signs of obstructed axonal transport have been reported in ischemia of the human eye [6].

The clinically observable correlate to obstructed axonal transport is though to be whitish areas of cotton-wool spots at the periphery of areas of retinal ischemia. Pale papilledema of peripapillary whitish areas reflecting similar axonal changes with interruption of anterograde or retrograde transport at the optic disc occur in ischemic optic neuropathy or after central artery occlusion [6].

Pathophysiological Role of Axonal Transport Blockade During Intraocular Pressure (IOP) Elevation

Acute IOP elevation causes obstruction of the rapid axonal transport at the optic nerve head in monkeys when the intraocular pressure is elevated to 25 to 50 mm Hg below mean arterial pressure [1, 12]. The site of the axonal transport blockade is at the scleral *lamina cribrosa,* where the retinal nerve fiber bundles pass through openings in the laminar sheets. Reversal of transport blockade occurred rapidly after normalization of intraocular pressure. In animals with increased intraocular pressure maintained at a perfusion pressure of 25 mm Hg Quigly and coworkers [13] could demonstrate an induced disturbance of axonal transport approximately after 3 h of compression. Following that time interval an increasing accumulation of labelled proteins was observed in the laminar region together with a 60% decrease of the rapid phase material in the retrobulbar optic pathway. This transport obstruction seems to be proportional to ocular perfusion pressure [9] and is caused by a blockade in a certain number of axons passing through the *lamina cribrosa.*

Electron microscopical and autoradiographical studies demonstrate that the transport obstruction is nonuniform with some axons accumulating organelles or radiolabelled material (preferentially at the upper and lower poles of the optic disc or the temporal quadrants of the optic nerve head) whereas other axons seem to be unaffected [1, 10, 12]. It is, therefore, probable that acute intraocular pressure elevation induces more or less complete rapid transport blockade in some axons, whereas no or only slight impairment of transport occurs in the remaining axons (Fig. 2a).

Retrograde axonal transport, as visualized by horseradish peroxidase transport, is also obstructed by moderate elevations of intraocular pressure as is the case with the anterograde transport [10]. Reversibility of the transport block has been demonstrated after moderate pressure blocks [10].

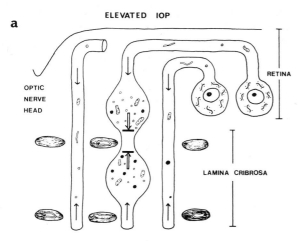

Fig. 2 a, b. Schematic drawing of the pattern of axonal transport obstruction at the optic nerve head in animals with elevated ocular pressure (a) or experimental disc swelling (b)

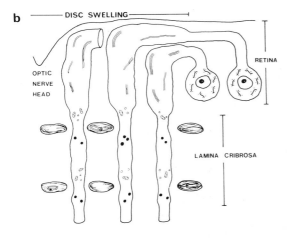

Comments

Even though the factors responsible for the transport obstruction in experimental acute intraocular pressure elevation are unknown, compressive insults or local tissue ischemia are the two hypotheses generally put forward [4].

The controversy between the ischemic and mechanical theories are not solved [4] although according to my opinion more observations favour that mechanical factors play a decisive role. For example, experiments where the IOP elevation was combined with inhalation of 100% oxygen or hyperbaric oxygen respiration the degree of axonal transport block was unchanged [10, 14]. Furthermore, study of cross sections of nerve bundles in monkey eyes subjected to acute IOP-elevation showed the greatest axonal transport abnormalities in the periphery of the nerve bundles, where the compressive or kinking effect on the axons is supposed to be maximal. In another study, however, a more diffuse transport blockade was observed within the axon bundles at the *lamina cribrosa* [15]. Such a transport interruption could be the result of a pressure gradient over the *lamina cribrosa,* where the axons pass from one compartment with high tissue pressure to one of lower pressure.

The alternative mechanism for axonal transport obstruction is microvascular collapse due to increased tissue pressure within the optic nerve head with subsequent local ischemia [4]. Recent experimental work, however, give little support to the conclusion that the optic nerve head circulation is severely compromised in acute IOP elevation [3, 10].

Axonal Transport Obstruction During Swelling of the Optic Nerve Head

Papilledema or swelling of the optic nerve head can be experimentally produced in monkeys by, for example, intracranial balloons or ocular hypotony [9, 17]. Electron-microscopy of the induced papilledema reveals that the major event in disc swelling is the appearance of swollen axons in the anterior part of the optic nerve head and the peripapillary retina. The morphological studies show that the interstitial edema is mild and does not account for any major part of the swelling of the optic disc. Axonal transport studies in experimental papilledema induced by elevated intracranial pressure [17] or ocular hypotony [9] demonstrated that the slow axonal transport is markedly obstructed within the optic nerve head. Interestingly enough, a similar axonal transport dysfunction is found in both experimental conditions even though the primary pathogenetic factors are different. Autoradiography demonstrates accumulation of slowly transported material within the whole optic nerve head anterior to the *lamina scleralis.* The appearance of marked axon swelling in papilledema is therefore generally attributed to the accumulation of intraaxonal material due to disturbance of the slow transport process (Fig. 2b). Since the major constituents of the slow phase are cytoskeletal proteins or proteins involved in motility, an accumulation of these proteins can be anticipated. The alterations of the rapid transport in papilledema seem to be less marked [9, 17] and the morphological counterpart of rapid phase obstruction, i.e. accumulation of organelles such as smooth surfaced vesicles and mitochondria, can only be seen focally at the *lamina cribrosa.* It is, therefore, probable that the major transport obstruction affects the slow phase; transport changes in the rapid phase are minor and possibly secondary to the slow phase obstruction. Experimental data also indicate that the obstruction of the slow transport has the form of a retardation or slowing-down of the slow phase when passing the optic nerve head (for discussion, see [2]) with the main part of the rapid transport remaining intact.

Summary

This short review has presented some recent data and theories concerning the implications of axonal transport studies in ophthalmology. The pattern of axonal transport block at the optic nerve head in experimental intraocular hypertension supports the view that mechanical compression at this site is a probable mechanism of optic nerve damage. Transport obstruction within the axons of primarily the slow phase seems to be a major factor in the production of the swollen axon in optic nerve head swelling. This new knowledge concerning axonal transport obstruction in retinal ganglion cell disorders has provided the ophthalmologist with a better base for interpretation of various clinical signs of disease observed in the human retina and optic disc.

Acknowledgments. Supported by the Swedish Medical Research Council (Grant No. 02226).

References

1. Anderson DR, Hendrickson A (1974) Effect of intraocular pressure on rapid axoplasmic transport in monkey optic nerve. Invest Ophthalmol 13:771–783
2. Anderson DR (1979) Papilledema and axonal transport. In: Thompson HS (ed) Topics in neuro-ophthalmology. Waverly Press, Baltimore, pp 184–189
3. Geijer C, Bill A (1979) Effects of raised intraocular pressure on retinal, prelaminar, laminar and retrolaminar optic nerve blood flow in monkeys. Invest Ophthalmol Vis Sci 18:1030–1042
4. Hayreh SS, March W (1979) Pathogenesis of block of rapid orthograde axonal transport by elevated intraocular pressure. Exp Eye Res 28:515–523
5. Holmgren E et al (1978) Changes in synaptic function induced by blockage of axonal transport in the rabbit optic pathway. Brain Res 157:267–276
6. McLeod D (1976) Ophthalmoscopic signs of obstructed axoplasmic transport after ocular vascular occlusions. Br J Ophthalmol 60:551–556
7. McLeod D, Marshall J, Kohner EM, Bird AC (1977) The role of axoplasmic transport in the pathogenesis of retinal cotton-wool spots. Br J Ophthalmol 61:177–191
8. McLeod D, Marchall J, Kohner EM (1980) Role of axoplasmic transport in the pathophysiology of ischaemic disc swelling. Br J Ophthalmol 64:247–261
9. Minckler DS et al. (1976) A light microscopic, autoradiographic study of axoplasmic transport in the optic nerve head during ocular hypotony, increased intraocular pressure and papilledema. Am J Ophthalmol 82:741
10. Minckler DS (1977) Orthograde and retrograde axoplasmic transport during acute ocular hypertension in the monkey. Invest Ophthalmol Vis Sci 16:426–441
11. Ochs S (1974) Energy metabolism and supply of ~ P to the fast axoplasmic transport mechanism in nerve. Fed Proc 33:1049–1058
12. Quigley HA, Anderson DR (1976) The dynamics and location of axonal transport blockade by acute intraocular pressure elevation in primate optic nerve. Invest Ophthalmol 15:606–625
13. Quigley HA, Guy J, Anderson DR (1979) Blockade of rapid axonal transport. Arch Ophthalmol 97:525–531

14. Quigley HA. Flower RW, Addicks EM, McLeod DS (1980) The mechanism of optic nerve damage in experimental acute intraocular pressure elevation. Invest Ophthalmol Vis Sci 19:505–517
15. Radius RL, Anderson DR (1980) Fast axonal transport in early experimental disc edema. Invest Ophthalmol Vis Sci 19:158–168
16. Rydevik B, McLean WG, Sjöstrand J, Lundborg G (1980) Blockade of axonal transport induced by acute, graded compression of the rabbit vagus nerve. J Neurol Neurosurg Psychiatry 43:690–698
17. Tso MOM, Hayreh SS (1977) Optic disc edema in raised intracranial pressure. Arch Ophthalmol 95:1458–1462

Implications of Axoplasmic Transport for the Spread of Virus Infections in the Nervous System

KRISTER KRISTENSSON [1]

In several cell systems viruses have provided a valuable tool for the study of transport and localization of newly synthesized proteins in a cell [12]. In the nervous system studies of the spread of viruses predicted the occurrence of a fast axoplasmic transport in both the retrograde and the anterograde direction [3, 32].

Retrograde Transport of Viruses

It is an old observation that viruses, such as the rabies virus, appear to spread along peripheral nerves to the central nervous system from an inoculation site at the periphery [29]. The pathways for the virus spread within the nerve were much debated. From experiments with herpes simplex virus (HSV) infection in rabbits Marinesco and Draganesco [24] reached the conclusion that the virus spread via a series of multiplications in endoneurial cells, while Goodpasture [11] using a similar model, suggested instead a spread within the axons. However, after Weiss and Hiscoe [37] had described the slow axonal flow in the somatofugal direction it was considered inconceivable that viral particles could travel in an opposite direction [39]. When virus antigen was seen in Schwann cells with immunofluorescence it was again suggested that the route for virus spread was via the Schwann cells and endoneurial tissue spaces [15, 38]. In a series of experiments designed to study the interaction between a neuron and HSV in vivo we obtained data indicating a rapid retrograde transport of HVS in nerves and that the axon indeed was the compartment [18]. After intradermal HSV inoculation the early lesions appeared to follow sensory pathways, while after intramuscular injection early lesions could be seen in motor neurons as well. Crushing or freezing the nerve prevented virus spread, as did topical application of colchicine and vinblastine to the nerve [19]. With the use of immunoperoxidase staining we also recently found an early labelling of the nerve cell bodies in the trigeminal ganglion after inoculation in the snout, where densely innervated areas of the vibrissae root sheaths are infected [23]. Evidence for a somatopetal transfer of viruses in axons has

1 Department of Pathology, Division of Neuropathology, Karolinska Institutet, Huddinge University Hospital, S-141 86 Huddinge, Sweden

Axoplasmic Transport in Physiology and Pathology
(ed. by D.G. Weiss and A. Gorio)
© Springer-Verlag Berlin Heidelberg 1982

also been obtained in other experimental models employing HSV [1, 6, 31] as well as rabies and pseudorabies viruses [2, 8, 25, 26, 36] and poliovirus [16]. Recently Ziegler and Herman [40] in a specially devised system in vitro have made observations consistent with the view that a virus is incorporated at axon terminals and transferred in the axons to the nerve cell bodies. In their system, neurites from sensory ganglion cells grow out from a cloning cylinder into a second compartment in Petri dishes. There is no diffusion of substances from the compartment with neurites alone to that containing the cell bodies. In spite of this, the cell bodies were infected when HSV was applied to the neurite compartment. In such a system the pricise mechanisms for virus uptake at the axonal surface, i.e. by endocytosis or fusion of the virus envelope with the axolemma [20], and the state in which the virus is travelling in the axon [8] can be examined.

Anterograde Transport of Viruses

There is also evidence for a fast axonal transport of viruses in the anterograde direction. When Goodpasture's [11] experiment of intraocular HSV injection was repeated, we found virus replication in ganglion cell bodies of the retina. Within 24 h infected neurons were also seen in the lateral geniculate body and superior colliculus on the contralateral side, which indicates that the virus had travelled with fast axoplasmic transport and passed to the postsynaptic neuron [21]. Herpes virus nucleocapsids are formed in the nucleus and bud through the inner nuclear membrane into the perinuclear space, thereby acquiring an envelope (Fig. 1). From this area they are trans-

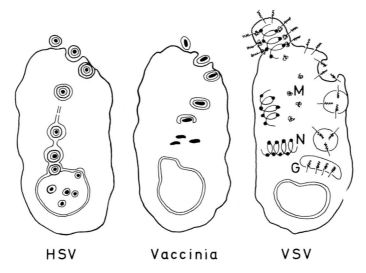

HSV   Vaccinia   VSV

Fig. 1. Examples of three different release mechanisms of viruses from a cell. *G* glycoprotein, *M* matrix protein, *N* nucleocapsid protein

ferred within endoplasmic reticulum (ER) membranes to the cell surface where they
are released [33]. Also in neuron soma, HSV can be seen within cisterns of ER and
occasionally the viral particles have been seen within cisterns or tubular structures in
axons [6, 14, 21]. It may therefore be suggested that the viruses utilize the smooth
ER membrane in the axons for their transport to the periphery. After its axonal
transport HSV can infect the postsynaptic neuron, but the exact mechanism for
release and uptake of virus at the synaptic area has not been evaluated. In this way
a chain of several neurons can be infected. For instance, in mice we have recently
observed that HSV antigen appears in the trigeminal ganglion cells after injection into
the snout. Later the second order of neuron in the main sensory and spinal tract
nuclei in the brainstem are infected. The third order of neurons are then infected,
which includes neurons in the reticular formation, the raphe nuclei, and the contralateral thalamus. Also, very conspicuously, neurons in the *locus coeruleus* on both
sides are infected, indicating that the monoamine producing neurons in the brainstem
may be readily accessible to this common virus [23] (Fig. 2). HSV remains latent in
trigeminal ganglia between recurrences of herpetic lesions [17]. When activated, HSV
appears to travel along peripheral branches of axons in the trigeminal nerve to the lip
or conjunctiva. It is not known whether the virus, at the same time, is also transported
in the central branches of the axons towards the brainstem in man; in our mouse
model with latent infection of trigeminal ganglion, a mild inflammation and demyelination occurred in the central part of the trigeminal root [22].

Paramyxo- and rhabdo-viruses, like measles and vesicular stomatitis viruses, use
quite different means for assembly, transport and release in cells [12, 34]. The glycoproteins of their future envelopes are synthesized in ER membranes and transferred
with them to the plasma membrane, which they reach by exocytosis (Fig. 1). The
matrix and nucleocapsid proteins instead are synthesized at free polyribosomes and
transferred solubilized in the cytoplasm to the area of the plasma membrane from

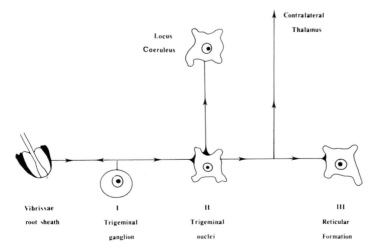

Fig. 2. Proposed pathways for axonal transport of herpes simplex virus in the trigeminal system
to monoaminergic brainstem nuclei after inoculation into the mouse snout

which the viral particles bud off. The matrix protein appears to make possible a recognition between the viral glycoprotein and nucleocapsid protein. Such a mode for virus protein transport and viral assembly in a neuron would imply that the glycoproteins are transferred in smooth ER membranes at a fast phase of transport in the axons, while the matrix and nucleocapsid proteins move at slow rates. With the help of monoclonal antibodies to the different proteins it should be possible to test whether such a hypothesis is valid [27]. Dubois et al. [9, 10] in a series of studies have demonstrated that in vitro the vesicular stomatitis virus can bud off from most of the nerve cell membrane. However, if the tissue cultures are treated with antibodies the viruses bud preferentially at synaptic areas. The viral particles were always found at the postsynaptic site and from here appeared to bud into the presynaptic ending. A similar budding from postsynaptic sites has also been described for rabies virus in vivo [5]. A remarkable localization of measles antigen to postsynaptic densities without any evidence of viral assembly has been observed in adult BALBlc mice infected with a hamster neurotropic strain of measles [30]. The reason for such selective budding or localization of virus antigen may reflect the structural differentiation of a neuron and due to this, the neurons normal transport of glycoproteins to different surface regions. In epithelial monolayer cells from kidney, for instance, it has been shown that while vesicular stomatitis viruses only bud off from the basolateral plasma membrane, Sendai virus and influenza virus bud only from the free, apical surface. Differences in contents of sialic acid in the viral glycoprotein between the viruses have been suggested to explain such differences, since the distribution of the sialic acid between the apical and the basolateral surface differs in normal kidney monolayer cells [4].

A third mechanism by which a virus may be released from a cell is that which has been observed for vaccinia virus. These viral particles are synthesized in the cytoplasm and are then surrounded by Golgi or ER membranes. Enwrapped within such a double layered membrane, the virus is transferred to the cell surface where the outer membrane fuses with the plasma membrane and the enveloped virus particle is released by exocytosis [28]. Such a release mechanism has so far not been studied in neurons (Fig. 1).

Implications of the Findings

The capacity of viruses to use axons for their transport may explain why viruses can spread over long distances to and within the nervous system and attack specific groups of neurons. For instance, the selective involvement of lower motor neurons in poliomyelitis may be due to an axonal transport in motor neuron axons following seeding of viruses into skeletal muscles during viremia. In fact, a tracer substance like HRP injected intravenously can diffuse into skeletal muscles, be incorporated by motor nerve terminals and selectively be transported to the lower motor neurons, and thereby bypass the blood-brain barrier. A slow axonal transport of viral proteins solubilized in the cytoplasm may explain why certain viruses, such as rabies, may have very long incubation periods that may extend over several months. Differential transport of different viral components in such highly specialized cells as neurons may also imply that the production of defective viral infections may be facilitated in the nervous system. A defect in production of matrix protein has been suggested to explain why measles virus occasionally causes the slowly progressive disease SSPE (subacute sclerosing panencephalitis) characterized by a defective budding of the virus from neurons [35]. If viral proteins from such defective virus infections are localized to different areas at synapses they might well alter the function of otherwise structurally intact neurons. Finally, axonal transport may be a way to specifically deliver certain antiviral agents to neurons. For instance, as HSV remains latent in trigeminal or sacral ganglion cell bodies between eruptions, the topical application of an antiviral agent or nerve cell body destroying agent in a suitable carrier would be the ideal treatment for these infections [7, 13].

Acknowledgments. Supported by a grant from the Swedish Medical Research Council, project No. B81-12X-04480-07A.

References

1. Baringer JR, Griffith JF (1970) Experimental herpes simplex encephalitis: early neuro-pathologic changes. J Neuropathol Exp Neurol 29:89–104
2. Vijlenga G, Heaney T (1978) Post-exposure local treatment of mice infected with rabies with two axonal flow inhibitors, colchicine and vinblastine. J Gen Virol 39:381–385
3. Bodian D, Howe HA (1941) The rate of progression of poliomyelitis virus in nerve. Bull Johns Hopkins Hosp. 69:79–85
4. Boulan ER, Sabatini DD (1978) Asymmetric budding of viruses in epithelial monolayers. A model system for study of epithelial polarity. Proc Natl Acad Sci USA 75:5071–5075
5. Charlton KM, Casey GA (1979) Experimental rabies in skunks. Immunofluorescence light and electron microscopic studies. Lab Invest 41:36–44
6. Cook ML, Stevens JG (1973) Pathogenesis of herpetic neuritis and ganglionitis in mice: evidence for intra-axonal transport of infection. Infect Immun 7:272–288
7. Desmukh DS, Kristensson K, Wisniewski HM, Brockerhoff H (1981) Toxicity and neuronal transport of stable liposomes and phospholipids in the nervous system. Neurochem Res 6:143–151
8. Dolivo M, Beretta E, Bonifas V, Foroglou C (1978) Ultrastructure and function in sympathetic ganglia isolated from rats infected with pseudorabies virus. Brain Res 140:111–125
9. Dubois-Dalcq M, Hooghe-Peters EL, Lazzarini RA (1980) Antibody induced modulation of rhabdovirus infection of neurons in vitro. J Neuropathol Exp Neurol 39:507–522
10. Faulkner G, Dubois-Dalcq M, Hooghe-Peters E, McFarland HF, Lazzarini RA (1979) Defective interfering particles modulate VSV infection of dissociate neuron cultures. Cell 17: 979–991
11. Goodpasture EW (1925) The axis-cylinders of peripheral nerves as portals of entry to the central nervous system for the virus of herpes simplex in experimentally infected rabbits. Am J Pathol 1:11–28
12. Grimley PM, Demsey A (1980) Functions and alterations of cell membranes during active virus infection. In: Trumph B, Arstila A (eds) Pathobiology of cell membranes, vol II. Academic Press, London New York, pp 93–168
13. Haschke RH, Ordrouneau JM, Bunt AH (1980) Preparation and retrograde axonal transport of an antiviral drug/horseradish peroxidase conjugate. J Neurochem 35:1431–1435

14. Hill TJ, Field HJ, Roome APC (1972) Intra-axonal location of herpes simplex virus particles. J Gen Virol 15:253–255
15. Johnson RT (1964) The pathogenesis of herpes encephalitis. I. Virus pathway to the nervous syssstem of suckling mice demonstrated by fluorescent antobody. J Exp Med 119:343–356
16. Jubelt B, Narayan O, Johnson RT (1980) Pathogenesis of human poliovirus infection in mice. II. Age-dependency of paralysis. J Neuropathol Exp Neurol 39:149–159
17. Klein RJ (1976) Pathogenetic mechanisms of recurrent herpes simplex virus infections. Arch Virol 51:1–13
18. Kristensson K (1970) Morphological studies of the neural spread of herpes simplex virus to the central nervous system. Acta Neuropathol 16:54–63
19. Kristensson K, Lycke E, Sjöstrand J (1971) Spread of herpes simplex virus in peripheral nerves. Acta Neuropathol 17:44–53
20. Kristensson K, Sheppard RP, Bornstein MB (1974) Observations on uptake of herpes simplex virus in organized cultures of mammalian nervous system. Acta Neuropathol 28:37–44
21. Kristensson K, Ghetti B, Wisniewski HM (1974) Study on the propagation of herpes simplex virus (type 2) into the brain after intraocular injection. Brain Res 69:189–201
22. Kristensson K, Svennerholm B, Persson L, Vahlne A, Lycke E (1979) Latent herpes simplex virus trigeminal ganglionic infection in mice and demyelination in the central nervous system. J Neurol Sci 43:253–264
23. Kristensson K, Nennesmo I, Persson L, Lycke E (1982) Neuron to neuron transmission of herpes simplex virus: transport of virus from skin to brainstem nuclei. J Neurol Sci 54:149–156
24. Marinesco G, Draganesco S (1923) Recherches expérimentales sur le neurotropisme du virus herpétique. Ann Inst Pasteur 37:752–783
25. McCracken RM, McFerran JB, Dow C (1973) The neural spread of pseudorabies virus in calves. J Gen Virol 20:17–28
26. Murphy FA, Harrison AK, Winn WC, Bauer SP (1973) Comparative pathogenesis of rabies and rabies-like viruses. Lab Invest 29–1–16
27. Norrby E, Swoveland P, Kristensson K, Johnson KP (1980) Further studies on subacute encephalitis and hydrocephalus in hamsters caused by measles virus from persistently infected cell cultures. J Med Virol 5:109–116
28. Payne L, Kristensson K (1979) Mechanism of vaccinia virus release and its specific inhibition by N-isonicotinoyl-N_2-3-methyl-4-chlorobenzoylhydrazine. J Virol 32:614–622
29. Pasteur L, Roux E, Chamberland C (1884) Nouvelle communication sur la rage. CR Acad Sci 98:457–463
30. Pottelsberghe C van, Rammohan KW, McFarland HF, Dubois-Dalcq M (1979) Selective neuronal, dendritic, and postsynaptic localization of viral antigen in measles-infected mice. Lab Invest 40:99–108
31. Price RW, Schmitz J (1979) Route of infection, systemic host resistance, and integrity of ganglionic axons influence acute and latent herpes simplex virus infection of the superior cervical ganglion. Infect Immun 23:373–383
32. Sabin AB (1941) Constitutional barriers to involvement of the nervous system by certain viruses, with special reference to the role of nutrition. J Pediatr 19:596–607
33. Schwartz J, Roizman B (1969) Concerning the egress of herpes simplex virus from infected cells: electron and light microscope observations. Virology 38:42–49
34. Simons K, Garoff H (1980) The budding mechanisms of enveloped animal viruses. J Gen Virol 50:1–21
35. Thormar H (1980) Subacute sclerosing panencephalitis. In: Neurochemistry and clinical neurology. Alan Liss, New York, pp 91–100
36. Tsiang H (1979) Evidence for an intraaxonal transport of fixed and street rabies virus. J Neuropathol Exp Neurol 38:286–299
37. Weiss PA, Hiscoe H (1948) Experiments on the mechanism of nerve growth. J Exp Zool 107:315–396
38. Wildy P (1967) The progression of herpes simplex virus to the central nervous system of the mouse. J Hyg 65:173–192
39. Wright GP (1953) Nerve trunks as pathways in infection. Proc R Soc Med 46:319–330
40. Ziegler RJ, Herman RE (1980) Peripheral infection in culture of rat sensory neurons by herpes simplex virus. Infect Immun 28:620–623

Section 5 Axoplasmic Transport as a Tool in Neurophysiology and Neuroanatomy

This section is meant as an introduction to some of the techniques which make use of axoplasmic transport. It was mainly due to the development of these techniques that neuroanatomy has made such extraordinary progress during the last decade, a progress which can only be compared to the advancement achieved by Cajal and Golgi when they added a new dimension to our understanding of the nervous system.

The chapters in this section survey the possibilities these methods offer to those studying the circuitry of the nervous system by tracing neural connections or by delineating the geometry of the cells from which one has recorded. Practical hints on the techniques itself cannot be given here because so many recipes are now known that they fill books themselves [1—5]. It is revealing that the first book on axoplasmic transport was one which dealt with its application to fiber tracing [1].

References

1. Cowan WM, Cuénod M (eds) (1975) The use of axonal transport for studies of neuronal connectivity. Elsevier, Amsterdam
2. Heimer L, RoBards MJ (eds) (1981) Neuroanatomical tract-tracing methods. Plenum Press, New York
3. Mesulam M-M (ed) (1982) Tracing neural connections with horseradish peroxidase. John Wiley & Sons, Chichester
4. Palay SL, Chan-Palay V (eds) Cytochemical methods in neuroanatomy. Alan Liss, New York (in press)
5. Robertson RT (ed) (1978) Neuroanatomical research techniques. Academic Press, New York

Retrograde Migration of Transmitter-Related Molecules

M. CUÉNOD[1], P. BAGNOLI[1,2], A. BEAUDET[1,3], A. RUSTIONI[1,4],
L. WIKLUND[1], and P. STREIT[1]

The present work is based on the general hypothesis that some transmitter or related molecules, after having been selectively taken up by the nerve terminals in which they are normally utilized, migrate (possibly in a modified form) toward the perikaryon where they accumulate. Selective uptake has been documented for the amino acid transmitters, for choline and for the biogenic amines [23, 25, 32] and somatopetal migration of material in the axon is a well established phenomenon [21]. The procedure presented here could prove to be both reliable and easy for delineating pathways, particularly those using amino acids or acetylcholine as transmitters.

Glycine

There is good evidence that glycine, which is well established as an inhibitory transmitter [1], could be involved in the pigeon isthmo-tectal neurons, as indicated by in vivo release experiments [29, 45]. ^{3}H-glycine, applied to the optic tectum led to perikaryal labeling of a small group of neurons within the *nucleus isthmi, pars parvocellularis* (Ipc) as shown by Hunt et al. [24] and Streit et al. [39], at both the light and electronmicroscopic levels. This pattern is best explained by assuming an intraaxonal retrograde migration of a radioactive compound. The retrograde labeling was observed after survival times as short as 20 min, indicating a speed of migration of the order of 150–250 mm per day, which is well compatible with that of retrograde axonal transport [39]. Under these conditions, a high density of silver grains was detected over the perikarya when glutaraldehyde was used as fixative while only faint labeling was observed after formaldehyde perfusion. This suggests that most of the radioactivity reaching Ipc neurons is in a soluble form [14, 39].

1   Brain Research Institute, University of Zurich, CH-8029 Zurich, Switzerland
2   Istituto di Fisiologia dell'Università di Pisa e Neurofisiologia del CNR, Via S. Zeno 31,
    I-56100 Pisa, Italy
3   Laboratory of Neuroanatomy, Montreal Neurological Institute, McGill University,
    3801 University Street, Montreal, Qué H3A 3B4, Canada
4   The University of North Carolina, Department of Anatomy, 111 Swing Bldg. H, Chapel Hill,
    NC 27514, USA

Axoplasmic Transport in Physiology and Pathology
(ed. by D.G. Weiss and A. Gorio)
© Springer-Verlag Berlin Heidelberg 1982

Gamma-Aminobutyric Acid

GABA is well established as transmitter in a population of striatal neurons projecting to the *substantia nigra*. After ^{3}H-GABA injection in the rat *substantia nigra,* Streit [42, 44] observed labeled fibers leading to the *caudoputamen,* where medium-sized neuronal perikarya are overlaid with silver grains. Perikaryal labeling was not observed in projections to the *substantia nigra* involving other transmitters, such as the ones originating in cerebral cortex or *nucleus raphé dorsalis.* Injections of ^{3}H–GABA in the dorsal horn of the cervical spinal cord led to intense labeling of about 12% of the bipolar neurons, exclusively large ones, in the spinal ganglion of the corresponding segment [31], an observation difficult to explain, because no evidence is available to date of GABAergic primary afferences.

D-Aspartate

The amino acids L-gluatmate (glu) and L-aspartate (asp) have been proposed as excitatory transmitters in the vertebrate central nervous system [11, 17, 26]. D-asp, which is not metabolized, acts as a false transmitter and is taken up by the same high affinity mechanisms as the L-forms of asp and glu [4, 18, 34, 35, 40]. Thus, ^{3}H-D-asp has been used to retrogradely label pathways presumably involving these excitatory amino acids as transmitters.

Many lines of evidence point to a transmitter role of glu and asp in retino-tectal neurons of the pigeon [13, 14, 22]. Following ^{3}H-D-asp application to the zone of termination of the optic nerve fibers in the optic lobe, Beaudet et al. [8, 15] could trace labeled axons up to the fiber layer of the contralateral retina, in which some ganglion cells were covered with silver grains. This labeling pattern suggests that ^{3}H-D-asp is selectively migrating within a subpopulation of retinal ganglion cells. Conversely, intravitreal injections of ^{3}H-D-asp led to labeling of a small percentage of perikarya in the ganglion cell layer [8, 20] and from there anterograde filling of the axons could be detected, leading to the contralateral tectum. Furthermore, the radioactivity transported anterogradely to the tectum belongs to a readily releaseable pool [8].

After ^{3}H-D-asp was injected, either in the dorsal horn at cervical levels or in the cuneate nucleus, Rustioni et al. [31] observed retrograde labeling of the afferent neurons in the corresponding spinal ganglia. The labeled cells were large and amounted to about 5% of the neuronal population in the ganglion.

In the cerebellum, after injections of ^{3}H-D-asp into various parts of the vermis, hemispheres and deep nuclei, retrograde labeling of the olivocerebellar climbing fiber pathway was observed by Wiklund et al. [44], while most parallel fibers and granule cells appeared unlabeled by the tracer. No mossy fiber afferents from brain stem or spinal cord were labeled by ^{3}H-D-asp [44].

Evidence is accumulating suggesting that many corticofugal neurons use glu/asp as transmitter. Application of ^{3}H-D-asp to the zone of termination of many of these corticofugal projections has led to their selective retrograde labeling. Injections in the striatum [36, 37], in the *substantia nigra* [36], in the lateral geniculate nucleus [7], in the dorsal column nuclei [30] and possibly in the dorsal horn of the cervical spinal cord [30] led to perikaryal labeling in the appropriate cortical areas.

Thus, in many systems, ^{3}H-D-asp labels pathways in a selective manner, which probably correspond to their use of asp or glu as transmitter.

Cholinergic Neurons

Cholinergic neurons and terminals pick up choline by a high affinity transport mechanism [6, 19, 27, 43, 47]. Bagnoli et al. [5] observed that ^{3}H-choline can be used to trace retrogradely cholinergic neurons in a selective way, provided the brains are unfixed, sectioned frozen and processed according to a dry mount autoradiograpic procedure. After ^{3}H-choline injections in the rat hippocampus, diffuse labeling, most likely anterograde, is observed in the contralateral hippocampus, in the ipsilateral entorhinal cortex, and in the lateral septum. Perikaryal foci of silver grains are present in the ipsilateral medial septal nucleus and nucleus of the diagnonal band, suggesting a retrograde labeling of this well established cholinergic pathway. In the pigeon, the thalamo-Wulst projection which is likely to contain cholinergic neurons [12, 41, 42] is also retrogradely labeled after ^{3}H-choline injections in the Wulst [5].

Biogenic Amines: Dopamine and Serotonin

Dopamine labels retrogradely the well established dopaminergic pathways originating in the *substantia nigra,* or in groups A8 and A10 and projecting to the *caudoputamen* [36]. However, it also labels cells in the dorsal raphé nucleus.

Leger et al. [28], Streit et al. [36, 38] and Beaudet et al. [9] reported retrograde perikaryal labeling in the *nucleus raphé dorsalis,* in the *substantia nigra* and in groups A10 and A8 after the injection of ^{3}H-serotonin into the rat *caudoputamen.* ^{3}H-Serotonin injections in the rat *substantia nigra* led to perikaryal labeling in the ipsilateral *nucleus raphé dorsalis* and occasionally in the cortex but not in the *caudoputamen* [36]. Superfusion of rat cerebral cortex with 10^{-4} M ^{3}H-serotonin induced retrograde perikaryal labeling in *nucleus raphé dorsalis, raphé medialis* and *locus coeruleus* [9]. Araneda et al. [2, 3] demonstrated retrograde labeling in the dorsal raphé nucleus after ^{3}H-serotonin injections in the olfactory bulb in rats pretreated with a monoaminoxidase inhibitor. Yamamoto et al. [46] injected ^{3}H-serotonin in the rat cerebellum and observed not only a perikaryal labeling of the raphé nuclei but also of many other non serotonergic brain stem nuclei known to project to the cerebellum.

Thus, serotonin retrograde labeling raises some problems: not only does it seem to cross label catecholamine neurons, as it might be expected from the established crossed specificity of their uptake mechanisms [10, 32], but it appears to become, under some circumstances, an almost unspecific marker which labels all the inputs to the injected area.

Conclusion

The data reviewed above fall into four classes based on the correlation between the results obtained with this approach and those available from other biochemical, histochemical, physiological and pharmacological established methodologies. The first two classes are double positive and double negative, the last two can be considered as "false negative" and "false positive".

There are pathways for which a given transmitter candidate is relative well established and which are retrogradely labeled after application of the appropriate marker, while other pathways are not known to use a given transmitter and are not retrogradely labeled after application of the corresponding marker. The body of evidence supplied by these two classes positively supports the hypothesis proposed.

The third class is made of the "false negative" results, namely cases in which a transmitter marker fails to label a group of neurons believed to release the corresponding transmitter: the failure to label GABAergic Purkinje neurons after ^{3}H-GABA injections in the Deiters nucleus (Wiklund, unpublished) is the prototype of this category. The fourth class collects the "false positive" results, that is to say those cases in which the application of a tritiated marker leads to retrograde perikaryal labeling of neurons for which the corresponding transmitter has not been suspected on the basis of other experimental approaches. This situation has been met among the biogenic amines, which usually crosslabel each other, and also with serotonin, which, under some circumstances, labels all afferences. The last two classes challenge the hypothesis proposed and most likely will have to be explained by reference to the uptake mechanism.

The chemical specificity seems related to uptake mechanisms. The retrograde labeling is observed only following microinjection of the tritiated substrate, but also after superfusion at concentrations in the range of high affinity. The radioactivity is intraaxonal [3, 39], and, after serotonin injections, the retrograde migration can be reduced by means of colchicine or vinblastine [2, 9]. The retrograde migration proceeds at about 150–200 mm/day for glycine [39] or 48 and 16 mm/day for serotonin [2, 3], rates fully compatible with that of retrograde transport of macromolecules [21]. The transmitter or its metabolite could then be bound to organelles such as vesicles, multivesicular bodies or lysosomes which are known to be transported retrogradely. Should a specific carrier be involved, it would follow that the specificity of the method does not rely only on the uptake mechanism, but also on a selective transport.

Summary

In conclusion, there is evidence favoring specific retrograde labeling of pathways according to their transmitter for amino acids, choline and biogenic amines. "False negative" and "false positive" observations were made in some pathways. The phenomenon seems to depend upon uptake mechanisms in the nerve terminal and a retrograde transport sensitive to tubulin-binding agents. Considering the limited amount of data available so far, the proposed hypothesis of a transmitter-specific retrograde labeling can by no means be considered as definitively established. The results are, however, promising enough to justify further investigation of this phenomenon, its mechanisms and its possible biological significance.

Acknowledgments. The authors wish to thank Ms. M. Jäckli and Ms. I. Roth for secretarial assistance. This work was supported by grants 3.505.79 and 3.506.79 of the Swiss National Science Foundation, the Dr. Eric Slack-Gyr-Foundation and Fellowships to A. Beaudet from the Roche Research Foundation for Scientific Exchange and the Medical Research Council of Canada.

References

1. Aprison MH, Nadi NS (1978) Glycine: Inhibition from the sacrum to the medulla. In: Fonnum F (ed) Amino acids as chemical transmitters. Plenum Press, New York, pp 531–570
2. Araneda S, Bobillier P, Buda M, Pujol J-F (1980) Retrograde axonal transport following injection of [^{3}H]serotonin in the olfactory bulb. I. Biochemical study. Brain Res 196:405–415
3. Araneda S, Gamrani H, Font C, Calas A, Pujol J-F, Bobillier P (1980) Retrograde axonal transport following injection of [^{3}H]serotonin into the olfactory bulb. II. Radioautographic study. Brain Res 196:417–427
4. Balcar VJ, Johnston GAR (1972) The structural specificity of the high affinity uptake of L-glutamate and L-aspartate by rat brain slices. J Neurochem 19:2657–2666
5. Bagnoli P, Beaudet A, Stella M, Cuénod M (1981) Selective retrograde labeling of cholinergic neurons with [^{3}H]choline. J Neurosci 1:691–695
6. Baughman RW, Bader CR (1977) Biochemical characterization of the cholinergic system in the chicken retina. Brain Res 138:469–485
7. Baughman RW, Gilbert CD (1981) Aspartate and glutamate as possible neurotransmitters in the visual cortex. J Neurosci 1:427–439
8. Beaudet A, Burkhalter A, Reubi JC, Cuénod M (1982) Selective bidirectional transport of [^{3}H]-D-aspartate in the pigeon retinotectal pathway. Neuroscience (in press)
9. Beaudet A, Stella M, Cuénod M (1982) On the specificity of the uptake and retrograde axonal transport of tritiated serotonin in rat central nervous system. Neuroscience 6:2021–2034
10. Berger B, Glowinski J (1978) Dopamine uptake in serotoningergic terminals in vitro: A valuable tool for the histochemical differentiation of catecholaminergic and serotininergic terminals in rat cerebral structures. Brain Res 147:29–45
11. Cotman CW, Foster A, Lanthorn T (1981) An overview of glutamate as a neurotransmitter. In: Di Chiara G, Gessa GL (eds) Glutamate as a neurotransmitter. Raven Press, New York, pp 1–27
12. Csoknay A, Vischer A, Bagnoli P, Henke H (1982) Cholinergic and GABAergic innervation in the pigeon visual Wulst: Effect of thalamic lesions. (In preparation)

13. Cuénod M, Henke H (1978) Neurotransmitters in the avian visual system. In: Fonnum F (ed) Amino acids as chemical transmitters. Plenum Press, New York, pp 221–239
14. Cuénod M, Streit P (1979) Amino acid transmitters and local circuitry in optic tectum. In: Schmitt FO, Worden FG (eds) The neurosciences: 4th study program. MIT Press, Cambridge (Mass) London, pp 989–1004
15. Cuénod M, Beaudet A, Canzek V, Streit P, Reubi JC (1981) Glutamatergic pathways in the pigeon and the rat brain. In: Di Chiara G, Gessa GL (eds) Glutamate as a neurotransmitter. Raven Press, New York, pp 57–68
16. Cuénod M, Bagnoli P, Beaudet A, Rustioni A, Wiklund L, Streit P (1982) Transmitter specific retrograde labeling of neurons. In: Chan-Palay V, Palay SL (eds) Cytochemical methods in neuroanatomy. Liss, New York, pp 17–44
17. Curtis DR, Johnston GAR (1974) Amino acid transmitters in the mammalian central nervous system. Ergeb Physiol 69:97–188
18. Davies LP, Johnston GAR (1976) Uptake and release of D- and L-aspartate by rat brain slices. J Neurochem 26:1007–1014
19. Dowdall MJ, Fox G, Wächtler K, Whittaker VP, Zimmermann H (1976) Recent studies on the comparative biochemistry of the cholinergic neuron. Cold Spring Harbor Symp Quant Biol 40:65–81
20. Ehinger B (1982) [³H]-D-aspartate accumulation in the retina of pigeon, guinea-pig and rabbit. Exp Eye Res 33:381–391
21. Grafstein B, Forman DS (1980) Intracellular transport in neurons. Physiol Rev 60:1167–1283
22. Henke H (1981) The central part of the avian visual system. In: Nistico G, Bolis L (eds) Progress in nonmammalian brain research. CRC Press, Boca Raton (USA) (in press)
23. Hökfelt T, Ljungdahl A (1975) Uptake mechanisms as a basis for the histochemical identification and tracing of transmitter-specific neuron populations. In: Cowan WM, Cuénod M (eds) Use of axonal transport for studies of neuronal connectivity. Elsevier, Amsterdam, pp 249–305
24. Hunt SP, Streit P, Künzle H, Cuénod M (1977) Characterization of the pigeon isthmo-tectal pathway by selective uptake and retrograde movement of radioactive compounds and by Golgi-like HRP. Brain Res 129:197–212
25. Iversen LL, Dick F, Kelly JS, Schon F (1975) Uptake and localisation of transmitter amino acids in the nervous system. In: Berl S, Clarke DD, Schneider D (eds) Metabolic compartmentation and neurotransmission. Plenum Press, New York, pp 65–87
26. Johnson JL (1978) The excitant amino acids glutamic and aspartic acid as transmitter candidates in the vertebrate central nervous system. Prog Neurobiol 10:155–202
27. Jope RS (1979) High affinity choline transport and acetylCoA production in brain and their roles in the regulation of acetylcholine synthesis. Brain Res 1:313–345
28. Leger L, Pujol JF, Bobillier P, Jouvet M (1977) Transport axoplasmique de la sérotonine par voie retrograde dans les neurones monoaminergiques centraux. CR Acad Sci 285:1179–1182
29. Reubi JC, Cuénod M (1976) Release of exogenous glycine in the pigeon optic tectum during stimulation of a midbrain nucleus. Brain Res 112:347–361
30. Rustioni A, Cuénod M (1982) Selective retrograde transport of D-aspartate in spinal interneurons and cortical neurons of rats. Brain Res 236:143–155
31. Rustioni A, Henke H, Cuénod M (1982) D-aspartate and GABA retrograde labeling of spinal ganglion neurons. (In preparation)
32. Shaskan E. Snyder SH (1970) Kinetics of serotonin accumulation into slices from rat brain: relationship to catecholamine uptake. J Pharmacol Exp Ther 175:404–418
33. Snyder SH, Yamamura HI, Pert CB, Logan WJ, Bennett JP (1973) Neuronal uptake of neurotransmitters and their precursors in studies with "transmitters" amino acids and choline. In: Mandell AJ (ed) New concepts in neurotrransmitter regulation. Plenum Press, New York, pp 195–222
34. Storm-Mathisen J (1979) Autoradiographic and microchemical localization of high affinity uptake of glutamate and aspartate. Proc Int Soc Neurochem (Jerusalem) 7:109
35. Storm-Mathisen J, Woxen Opsahl H (1978) Aspartate and/or glutamate may be transmitters in hippocampal efferents to septum and hypothalamus. Neurosci Lett 9:65–70

36. Streit P (1980) Selective retrograde labeling indicating the transmitter of neuronal pathways. J Comp Neurol 191:429–463
37. Streit P, Cuénod M (1979) Transmitter specificity and connectivity revealed by differential retrograde labeling of neural pathways. Neurosci Lett Suppl 3:340
38. Streit P, Knecht E, Cuénod M (1979) Transmitter specific retrograde labeling in the striato-nigral and raphe-nigral pathways. Science 205:306–308
39. Streit P, Knecht E, Cuénod M (1980) Transmitter related retrograde labeling in the pigeon optic lobe; a high resolution autoradiographic study. Brain Res 187:59–67
40. Takagaki G (1977) Properties of the accumulation of D-[^{14}C]aspartate into rat cerebral crude synaptosomal fraction. In: Fonnum F (ed) Amino acids as chemical transmitters. Plenum Press, New York, pp 357–361
41. Vischer A, Fäh A, Burkhalter A, Henke H (1980) Kainic acid toxicity in the pigeon thalamus and consequent decrease in the hyperstriatal choline acetyltransferase and glutamic acid decarboxylase. Experientia 36:703
42. Vischer A, Cuénod M, Henke H (1982) Neurotransmitter receptor ligand binding and enzyme regional distribution in the pigeon visual system. J Neurochem (in press)
43. Whittaker VP, Dowdall MJ (1975) Current state of research on cholinergic synapses. In: Waser PG (ed) Cholinergic mechanisms. Raven Press, New York, pp 35–40
44. Wiklund L, Toggenburger G, Cuénod M (1972) Aspartate: Possible neurotransmitter in cere-bellar climbing fibers. Science 216:78–80
45. Wolfensberger M, Reubi JC, Canzek V, Redweik U, Curtius HC, Cuénod M (1981) Mass frag-mentographic determination of endogenous glycine and glutamic acid released in vivo from the pigeon optic tectum. Effect of electrical stimulation of a midbrain nucleus. Brain Res 224:327–336
46. Yamamoto M, Chan-Palay V, Palay SL (1980) Autoradiographic experiments to examine uptake, anterograde and retrograde transport of tritiated serotonin in the mammalian brain. Anat Embryol 159:137–149
47. Yamamura HI, Snyder SH (1973) High affinity transport of choline into synaptosomes of rat brain. J Neurochem 21:1355–1374

Intra-Axonal Transport of Horseradish Peroxidase (HRP) and Its Use in Neuroanatomy

ANDREAS ASCHOFF and KLAUS SCHÖNITZER [1]

The intra-axonal transport (IT) of HRP was first demonstrated by Kristensson and Olsson [46] in the hypoglossal nerve of the rat and rabbit. Later it was suggested that the retrograde transport (RT) [47, 48], anterograde transport (AT) [55] or transneuronal transport [63, 92] of HRP might be useful for tracing neuronal connections in the central nervous system (CNS). Since then numerous papers were published dealing with the IT of HRP mainly in connection with fiber tracing. Due to limitation of space we focussed on a few aspects of the HRP technique, and on some ultrastructural findings on its uptake and IT. Since several reviews exist [5, 23, 41, 44, 45, 47, 48, 82a, 93] we concentrate on recent literature only. When using HRP as a tracer for studying neuronal connectivity it must be taken into account that several factors might influence the specificity or sensitivity of anterograde or retrograde labeling. Such factors are: (1) The type of enzyme used; (2) factors effecting uptake and IT; (3) uptake into fibers of passage (FOP); (4) tissue fixation [1, 29, 42], and (5) histochemical technique.

The Transported Protein, HRP

HRP is a glycoprotein with a molecular weight of 40,000 D and a molecular radius of 3 nm. It is composed of more than 20 isoenzymes with isoelectric points ranging from 3.5 to 9.0. The isoenzymes with the highest peroxidase activity have isoelectric points between 5.5 and 7.6 [74]. It was shown that basic isoenzymes of HRP are more effective markers than acidic ones [9, 10], and that HRP samples from different suppliers can produce different results [1]. In the visual system of the frog, basic isoenzymes of HRP are transported both anterogradely and retrogradely, but acidic isoenzymes only anterogradely [21]. AT of basic and acidic isoenzymes was also observed in the crushed hypoglossal nerve of the mouse [68]. It is, however, not clear whether the isoelectric point of the HRP isoenzymes is the only determinant for uptake and/or IT.

---

1 Max-Planck-Institut für Psychiatrie, Kraepelinstr. 2, D-8000 München 40, Fed. Rep. of Germany

Axoplasmic Transport in Physiology and Pathology
(ed. by D.G. Weiss and A. Gorio)
© Springer-Verlag Berlin Heidelberg 1982

Some isoenzymes are taken up and incorporated into "synaptic" or coated vesicles but are not transported [9]. This implicates an unspecific uptake mechanism for HRP but a selective transport mechanism [5].

Determinants for Uptake and IT or HRP

Uptake of HRP into neurons and its IT is determined mainly by four factors: (1) The morphology of the application site, e.g. the terminal density; (2) the application procedure; (3) the activity of the neurons (see below), and (4) by agents acting on the neuronal membrane.

Influence of Terminal Distribution at the Application Site on Retrograde Labeling

The density of terminals at the application site is an important factor for the retrograde labeling with HRP [39, 53]. After injections of HRP into the cerebral cortex of rat, monkey and cat [42], neurons in the intralaminar nuclei were labelled more weakly than those of the main thalamic relay nuclei. It was shown that the terminals of the intralaminar neurons are diffusely distributed all over the frontal, medial, and parietal cortex, whereas the neurons of the main thalamic relay nuclei have dense terminal fields in distinct laminae of specific cortical fields [39, 40].

These findings may help to explain why, in some systems [31, 64], no retrograde labeling with HRP was demonstrated. It is possible that widely distributed terminals have only limited access to the HRP pool at a given application site and uptake and RT in such cases is too low to be detected with certain histochemical methods.

Influence of the Application Procedure on Retrograde Labeling

Hamilton syringes [41, 48, 79] which allow accurate measurements of the tracer volumes injected are usually used for injection of HRP. However, the needles that are supplied with the microliter syringes usually are quite thick (over 300 μm diameter) and cause extensive damage to the tissue [7, 79]. Leakage of tracer into the track of the needle therefore may lead to false positive labeling [27] since HRP is readily taken up by damaged axons (see below). This problem is largely avoided by injection of HRP with micropipettes, either by pressure or by iontophoresis (e.g. [7, 18, 26]). There are drawbacks with the iontophoresis of HRP [24, 30, 52, 57], e.g. the amount of tracer applied cannot be determined accurately [57], and uptake of HRP into FOP seems to be facilitated by the iontophoresis technique [30]. It was shown, however, that with iontophoretic application diffusion of HRP into the track of the pipette is largely avoided [66]. The pressure injection of HRP through micropipettes is considered to be no less than or even more suitable for the application of HRP, than the iontophoresis technique [41, 79].

For chronical application of HRP in high concentration implantation or injection of HRP pellets or HRP gels [24, 52, 62, 78, 90] or application of HRP in crystalline form [22, 61, 65, 67] is used. The use of acrylamide matrices for HRP [24] may also prevent the inflammatory reaction of the tissue [88] caused by HRP at the application site. In those cases, where the application site is either naturally or by preparation isolated from the rest of the nervous tissue, e.g. eye [13, 18, 21, 35, 49] or transected peripheral nerves [17, 18, 19, 65, 67], large amounts of tracer may be applied without the danger of diffusion.

Agents Acting on the Neuronal Membrane

The dilution of HRP caused by diffusion from the application site [24, 29, 53, 75, 81], sparse uptake of HRP into neurons [80], as well as the failure to label axonal terminals by AT (see [60]) has led to the search of different agents that enhance the neuronal uptake of HRP into neurons. Uptake of HRP is enhanced by adding poly-L-ornithine [25, 35], dimethyl sulfoxide (DMSO) [43, 53, 91], lysophosphatidyl-choline [18, 53], and various detergents including NONIDET P-40 [18, 53] to the HRP solution.

Uptake of HRP in Fibers of Passage (FOP)

As already mentioned, HRP is taken up by fibers running through the injection site. It may then lead to labeling of perikarya [27, 67, 91] or terminals [12, 60, 89] belonging to the FOP. Anterograde labeling by fibers running through the injection site was denied by other authors [29, 91]. La Vail et al. [50] stated that retrograde labeling of perikarya which belong to FOP occurs only if the fibers are cut or otherwise damaged. Massive uptake of HRP by unmyelinated fibers versus sparse uptake by myelinated fibers was reported by Lund et al. [54], and uptake into mechanically undamaged axons by Herkenham and Nauta [30]. Undamaged axons take up HRP, but the uptake and IT usually is so low that retrograde labeling does not occur [6, 72]. This indicated, however, that labeling of neurons by undamaged FOP may occur when the uptake rate of HRP into the axon is sufficiently high and the histochemical demonstration of the enzyme is sufficiently sensitive. The development of sensitive methods for the demonstration of HRP (e.g. with TMB: [59]) and the use of detergents enhancing the uptake of HRP into the neurons may thus lead to the detection of HRP in perikarya or terminals belonging to FOP.

Histological Procedures

The classical histochemical procedure is incubating the material (usually frozen sections or vibratome sections) with diaminobenzidine (DAB) and H_2O_2. There are many variations of this procedure, using different concentrations and buffers (e.g. [49, 56]), adding different substances (e.g. [1, 3, 83]) or producing H_2O_2 by a coupled oxidation reaction in the incubation soak [36]. Another approach was the use of different chromogens instead of DAB: Benzidine dihydrochloride (BDHC) (e.g. [16]), p-phenylenediamine dihydrochloride and pyrocatechol (PPD-PC) [28], o-dianisidine (O-D) [15], o-tolidine (O-T) [82], tetramethylbenzidine (TMB) [58] as well as several others (see reviews [41, 48, 82a]). The TMB method [58] with its modifications [2, 60] has proved to be superior to other methods [12, 59, 78, 89] and seems to be the most commonly used HRP method for light microscopy (LM). Although TMB was considered to be inadequate for electron microscopic (EM) investigations [2, 3, 6], it was shown to be well suited for retrograde and anterograde EM tracing [77, 78, 84]. The TMB reaction product forms crystal-like electron dense structures, which can easily be recognized. In contrary to TMB the other chromogens which were used for EM are amorphous: DAB, which was commonly used, PPD-PC [11, 76] and O-T [82]. Therefore, they may cause equivocal results [69, 76, 78] requiring control experiments (e.g. [48, 51]). To investigate the organelles of HRP uptake and/or transport or to achieve a Golgi-like staining implicates the use of an amorphous chromogen like DAB. Furthermore, DAB is better suited to judge the strong staining at injection sites [4].

Ultrastructural Findings on Uptake, Transport, and Deposition of HRP

Uptake

After an injection of HRP into the CNS the protein diffuses into all extracellular spaces near the injection site [9, 49, 73, 86] where it can be found for some hours. Similar extracellular diffusion is found in the retina after injection of HRP into the vitreous body [35] or after injection into peripheral terminal fields [37, 46, 71]. The uptake of HRP following its application may either be restricted to membrane bound organelles (see below) or result in a diffuse filling of the neurons and/or neuronal processes near the injection site, producing a Golgi-like staining. It has been suggested [6, 43, 49, 86] that such Golgi-like labeling is caused by a local damage of the plasma membrane and successive diffusion into the neuron. However, membrane damage is not always considered to be necessary for Golgi-like labeling [35, 88]. Golgi-like staining is also achieved after intracellular HRP application (see [87] for literature). It is believed that the diffuse HRP can be converted into membrane bound vesicles [35, 88]. The Golgi-like staining near an injection site is often used for combined LM and EM investigations on the morphology of specific neurons or axonal branches and their circuity (e.g. [33, 88]). Apart from the diffuse filling with HRP the protein is taken up by neurons and other cells by pinocytosis. Pinocytotic vesicles filled with HRP were seen in the CNS [6, 38, 49, 86] as well as in the peripheral nervous system [37] and in tissue culture [8]. HRP can be taken up by neurons as well as by glial cells [35, 86], Schwann cells [71], and ependymal cells [70]. Neurons incorporate HRP through perikaryal, dendritic, and axonal membranes [6, 49, 86]. On the sites of axonal terminals in the CNS [6, 38, 49, 73, 86] as well as sensory nerve endings [37] HRP is mainly taken up by "synaptic" vesicles and/or coated vesicles. Other organelles which are considered to be associated with HRP uptake are: multivesicular bodies [9, 85], tubular cisternae and vesicles [8, 9, 38], lysosomes [70], and cupshaped bodies [8]. The uptake of HRP by small pinocytotic vesicles depends on the synaptic or neurosecretory activity [6, 34]. This coincides with the model of "membrane recycling", which says that the release of transmitter by exocytosis is followed by endocytosis (see [34] for review and literature). Furthermore, this coincides with the finding that the uptake of HRP in axonal terminals is far more effective than the uptake at perikaryal and dendritic membranes of undamaged neurons [6].

Transport

The organelles which are reported to transport HRP retrogradely and anterogradely are vesicles of different size (though always larger than synaptic vesicles) [6, 36, 49, 83], lysosomal bodies [86], multivesicular bodies [10, 49, 86], cupshaped organelles [6, 9, 36, 49], and tubular structures [6, 8, 9, 13, 49, 73, 86]. It was suggested that these tubular structures might be part of the smooth endoplasmic reticulum (SER) [13, 49, 73, 83] which was postulated to be a continuous channel throughout the length of the axon ([17a], see also [51]). Other authors [6, 85] pointed out that the tubular structures which transport HRP in the retrograde direction are different from the SER and are probably not connected to it. A recent investigation [51] demonstrated this in the chick visual system. For the AT of HRP, however, involvement of the SER remains possible.

The rates for the anterograde and/or retrograde transport of HRP estimated by different authors vary considerably from 1 mm/day [55] to about 50–100 mm/day [29, 49, 73] or more [13] up to 288–432 mm/day [60]. The differences may partly be due to the different animals and pathways under investigation, or result from the different sensitivity of the histochemical procedures [60].

Deposition

After RT HRP is finally deposited in the perikarya in dense bodies ([6, 49, 73, 86] and others). These were generally assumed to be lysosomes, which was proved by Broadwell and Brightman [6]. The labeled lysosomes are often found close to the Golgi apparatus [36, 49, 73, 86]. HRP deposit in axonal terminals after AT is generally described as forming small vesicles and tubules [13, 14, 35, 36, 82, 83]. Furthermore, diffuse staining was described additional to the vesicular labeling [83] or instead of it, then resembling Golgi-like staining [32]. Sometimes anterogradely transported HRP was not visible in the LM though it was found with the EM [14, 49], possibly because the HRP labeled vesicles were below the resolution of the LM [14].

References

1. Adams JC (1977) Technical considerations on the use of HRP as a neuronal marker. Neuroscience 2:141–145
2. Adams JC (1980) Stabilizing and rapid thionin staining of TMB-based HRP reaction product. Neurosci Lett 17:7–9
3. Adams JC (1981) Heavy metal intensification of DAB-based HRP reaction product. J Histochem Cytochem 29:775
4. Aschoff A, Holländer H (1982) Fluorescent compounds as retrograde tracers compared with HRP. I. A parametric study in the central-visual system of the albino rat. J Neurosci Methods (in press)
5. Bisby MA (1980) Retrograde axonal transport. In: Fedoroff S, Hertz L (eds) Advances in cellular neurology. Academic Press, London New York, pp 69–117
6. Broadwell RD, Brightman MW (1979) Cytochemistry of undamaged neurons transporting exogenous protein in vivo. J Comp Neurol 185:31–74
7. Bullier J, Henry GH, Baker WJ (1980) A simple device for injection of small calibrated amounts of HRP into the cerebral cortex. J Neurosci Methods 2:47–49
8. Bunge MB (1977) Initial endocytosis of HRP or ferritin by growth cones of cultured nerve cells. J Neurocytol 6:407–439
9. Bunt AH, Haschke RH (1978) Features of foreign proteins affecting their retrograde transport in axons of the visual system. J Neurocytol 7:665–678
10. Bunt AH, Haschke RH, Lund RD, Calkins DF (1976) Factors affecting retrograde axonal transport of HRP in the visual system. Brain Res 102:152–155
11. Carson KA, Lucas WJ, Gregg JM, Hanker JS (1980) Facilitated ultracytochemical demonstration of retrograde axonal transport of HRP in peripheral nerve. Histochemistry 67: 113–124
12. Chronister RB, Sikes RW, Wood J, Defrance JF (1980) The pattern of termination of ventral tegmental afferents into *nucleus accumbens:* An anterograde HRP analysis. Neurosci Lett 17:231–235
13. Colman DR, Scalia F, Cabrales E (1976) Light and electron microscopic observations on the anterograde transport of HRP in the optic pathway in the mouse and rat. Brain Res 102: 156–163
14. Condé F, Condé H (1979) Observations on the orthograde and retrograde transport of HRP in the cat. J Hirnforsch 20:35–46
15. De Olmos J (1977) An improved HRP method for the study of central nervous system. Exp Brain Res 29:541–551
16. De Olmos J, Heimer L (1977) Mapping of collateral projections with the HRP-method. Neurosci Lett 6:107–114
17. Deuschl G, Illert M (1981) Cytoarchitectonic organization of lumbar preganglionic sympathetic neurons in the cat. J Autonom Nerv Syst 3:193–213
17a. Droz B, Rambourg A (1982) Axonal smooth endoplasmic reticulum and fast orthograde transport of membrane constituents. In: Weiss DG (ed) Axoplasmic transport. Springer, Berlin Heidelberg New York, pp 384–385
18. Frank E, Harris WA, Kennedy MB (1980) Lysophosphatidyl choline facilitates labeling of CNS projections with HRP. J Neurosci Methods 2:183–189
19. Fritz N, Illert M, Saggau P (1981) Location of dorsal interosseus motor nuclei in the cat. Neurosci Lett 21:243–248
20. Gallager WD, Pert A (1978) Afferents to brain stem nuclei (brain stem raphe, *nucleus reticularis pontis caudalis* and *nucleus gigantocellularis*) in the rat as demonstrated by microiontophoretically applied HRP. Brain Res 144:257–275
21. Giorgi PP, Zahnd J (1978) Anterograde and retrograde transport of HRP isoenzymes in the retino-tectal fibres of *Xenopus* larvae. Neurosci Lett 10:109–114
22. Gobel S, Falls WM (1979) Anatomical observations on HRP-filled terminal primary axonal arborization in layer II of the *substantia gelatinosa* of Rolando. Brain Res 175:335–340

23. Grafstein B, Forman DS (1980) Intracellular transport in neurons. Physiol Rev 60:1167–1283

24. Griffin G, Watkins LR, Mayer DJ (1979) HRP pellets and slow-release gels: Two new techniques for greater localization and sensitivity. Brain Res 168:595–601

25. Hadley RT, Trachtenberg MC (1978) Poly-L-ornithine enhances the uptake of HRP. Brain Res 158:1–14

26. Haight JR, Sanderson KJ, Neylon L, Patten GS (1980) Relationships of the visual cortex in the marsupial brush tailed possum, *Trichosorus vulpecula*, a HRP and autoradiographic study. J Anat 131:387–413

27. Halpein JJ, LaVail JH (1975) A study of the dynamics of retrograde transport of HRP in injured neurons. Brain Res 100:253–269

28. Hanker JS, Yates PE, Metz CB, Rustioni A (1977) A new specific, sensitive and non carcinogenic reagent for the demonstration of HRP. Histochem J 9:789–792

29. Hedreen JC, McGrath S (1977) Observations on labeling of neuronal cell bodies, axons and terminals after injection of HRP into rat brain. J Comp Neurol 176:225–246

30. Herkenham M, Nauta WJH (1977) Afferent connections to the habenular nuclei in the rat. A HRP study, with a note on the fiber-of-passage problem. J Comp Neurol 173:123–146

31. Holm P, Flindt-Egebak P (1976) The absence of HRP uptake by cerebellar afferents to the red nucleus of the cat. Neurosci Lett 2:315–318

32. Holstege JC, Dekker JJ (1979) Electron microscopic identification of mamillary body terminals in the rat's AV thalamic nucleus by means of anterograde transport of HRP. A quantitative comparison with the EM degeneration and EM autoradiographic techniques. Neurosci Lett 11:129–135

33. Holländer H, Vanegas H (1981) Identification of pericellular baskets in striate cortex: Light and electron microscopic observations after uptake of HRP. J Neurocytol 10:577–587

34. Holtzman E (1977) The origin and fate of secretory packages, especially synaptic vesicles. Neuroscience 2:327–355

35. Itaya SK, Williams TH, Engel EL (1978) Anterograde transport of HRP enhanced by poly-L-ornithine. Brain Res 150:170–176

36. Itoh K, Konishi A, Nomura S, Mizuno N, Nakamura Y, Sugimoto T (1979) Application of coupled oxidation reaction to electron microscopic demonstration of HRP: Cobalt glucose oxidase method. Brain Res 175:341–346

37. Jirmanova I, Zelena J (1980) Uptake of HRP by sensory terminals of lamellated corpuscles in mouse foot pads. Acta Neuropathol 52:129–135

38. Jones DG, Cameron PU, Ellison LT (1977) Uptake of HRP by cortical synapses in rat brain. Cell Tissue Res 178:355–373

39. Jones EG (1975) Possible determinants of the degree of retrograde neuronal labeling with HRP. Brain Res 85:249–253

40. Jones EG (1975) Lamination and differential distribution of thalamic afferents within the sensory motor cortex of the squirrel monkey. J Comp Neurol 160:167–207

41. Jones EG, Hartman BK (1978) Recent advances in neuroanatomical methodology. Annu Rev Neurosci 1:215–296

42. Jones EG, Leavitt RY (1974) Retrograde axonal transport and the demonstration of nonspecific projections to the cerebral cortex and striatum from thalamic intralaminar nuclei in the rat, cat and monkey. J Comp Neurol 154:349–378

43. Keefer DA (1978) HRP as a retrogradely transported detailed dendritic marker. Brain Res 140:15–32

44. Kristensson K (1975) Retrograde axonal transport of protein tracers. In: Cowan WM, Cuénod M (eds) The use of axonal transport for studies of neuronal connectivity. Elsevier, Amsterdam, pp 70–82

45. Kristensson K (1978) Retrograde transport of macromolecules in axons. Annu Rev Pharmacol Toxicol 18:97–110

46. Kristensson K, Olsson Y (1971) Uptake and retrograde transport of peroxidase in hypoglossal neurons. Acta Neuropathol 19:1–9

47. La Vail JH (1975) Retrograde cell degeneration and retrograde transport techniques. In: Cowan MW, Cuénod M (eds) The use of axonal transport for studies of neuronal connectivity. Elsevier, Amsterdam, pp 218–248

48. La Vail JH (1978) A review on the retrograde transport technique. In: Robertson RT (ed) Neuroanatomical research techniques. Academic Press, London New York, pp 355–384

49. La Vail JH, La Vail MM (1974) The retrograde intraaxonal transport of HRP in the chick visual system: A light and electron microscopic study. J Comp Neurol 157:303–358

50. La Vail JH, Winston KR, Tish A (1973) A method based on retrograde intraaxonal transport of protein for identification of cell bodies of origin of axons terminating within the CNS. Brain Res 58:470–477

51. La Vail JH, Rapisardi S, Sugino IK (1980) Evidence against the smooth endoplasmic reticulum as a continuous channel for the retrograde axonal transport of HRP. Brain Res 191: 3–20

52. Leichnetz GR (1981) The median subcallosal fasciculus in the monkey: A unique prefrontal corticostriate and corticocortical pathway revealed by anterogradely transported HRP. Neurosci Lett 21:137–142

53. Lipp HP, Schwegler H (1980) Improved transport of HRP after injection with a non ionic detergent (Nonidet P-40) into mouse cortex and observations on the relationship between spread at the injection site and amount of transported label. Neurosci Lett 20:49–54

54. Lund JS, Lund RD, Hendrickson AE, Bunt AH, Fuchs AE (1975) The origin of afferent pathways from the primary visual cortex, area 17, of the macaque monkey as shown by retrograde transport of HRP. J Comp Neurol 164:287–304

55. Lynch G, Gall C, Mensah P, Cotman C (1974) HRP histochemistry: A new method for tracing efferent projections in the central nervous system. Brain Res 65:373–380

56. Malmgren L, Olsson Y (1978) A sensitive method for histochemical demonstration of HRP in neurons following retrograde axonal transport. Brain Res 148:279–294

57. McCaman RE, McKenna DG, Ono JK (1977) A pressure system for intracellular and extracellular ejection of picoliter volumes. Brain Res 136:141–147

58. Mesulam MM (1978) Tetramethyl benzidine for HRP neurohistochemistry: A noncarcinogenic blue rection product with superior sensitivity for visualizing neural afferents and efferents. J Histochem Cytochem 26:106–117

59. Mesulam MM, Rosene DL (1979) Sensitivity in HRP neurohistochemistry: A comparative and quantitative study of nine methods. J Histochem Cytochem 27:763–773

60. Mesulam MM, Mufson EJ (1980) The rapid anterograde transport of HRP. Neuroscience 5: 1277–1286

61. Mori J, Hori N, Katsuda N (1981) A new method for application of HRP into restricted area of the brain. Brain Res Bull 6:19–22

62. Moskowitz M, Mayberg M, Langer RS (1981) Controlled release of HRP from polymers: A method to improve histochemical localization and sensitivity. Brain Res 212:460–465

63. Nassel DR (1981) Transneuronal labeling with HRP in the visual system of the house fly. Brain Res 206:431–438

64. Nauta HJW, Pritz MB, Lasek RJ (1974) Afferents to the rat caudoputamen studied with HRP. An evaluation of a retrograde neuroanatomical research method. Brain Res 67:219–238

65. Neuhuber W, Niederle B (1980) Differential labeling by HRP of small and large spinal neurons of rats. Neurosci Lett 20:131–134

66. Newman R, Winans SS (1980) An experimental study of the ventral striatum of the golden hamster. I. Neuronal connections of the *nucleus accumbens*. J Comp Neurol 191:167–192

67. Oldfield BJ, McLachaln EM (1980) The segmental origin of preganglionic axons in upper thoracic rami of the cat. Neurosci Lett 18:11–17

68. Olsson Y, Malmgren LT (1980) Axonal uptake of HRP isoenzymes during Wallerian degeneration. Neurosci Lett 20:135–140

69. Osculati F, Gazzanelli G, Marelli M, Franceschini F, Amati S, Cinti S (1980) Critical appraisal of the technique of labeling neurons by retrograde transport of HRP. J Submicrosc Cytol 12:391–400

70. Pelletier G, Dupon A, Puviani R (1975) Ultrastructural study of uptake of peroxidase by the rat median eminence. Cell Tissue Res 156:521–532

71. Persson LA, Kristensson K (1979) Uptake of HRP in sensory nerve terminals of mouse trigeminal nerve. Acta Neuropathol 46:191–196

72. Phillipson OT (1979) Afferent projections to the ventral tegmental area of *Tsai* and interfascicular nucleus: A HRP study in the rat. J Comp Neurol 187:117–144

73. Price P, Fisher AWF (1978) Electron microscopical study of retrograde axonal transport of HRP in the supraoptico-hypophysial tract in the rat. J Anat 125:137–147

74. Rennke HG, Venhatachalam MA (1979) Chemical modification of HRP. Preparation and characterization of tracer enzymes with different isoelectric points. J Histochem Cytochem 27:1352–1353

75. Riley JN, Marchand ER (1981) A new microelectrophoretic procedure for delivery of HRP. Brain Res 205:396–399

76. Riley JN, Card JP, Moore RY (1981) A retinal projection to the lateral hypothalamus in the rat. Cell Tissue Res 214:257–269

77. Sakumoto T, Nagai T, Kimura H, Maeda T (1980) Electron microscopic visualization of tetramethyl benzidine reaction product on HRP neurohistochemistry. Cell Mol Biol 26: 211–216

78. Schönitzer K, Holländer H (1982) Anterograde tracing of HRP with the electron microscope using the tetramethylbenzidine reaction. J Neurosci Meth 4:373–383

79. Schubert P, Holländer H (1975) Methods for delivery of tracers to the central nervous system. In: Cowan WM, Cuénod M (eds) The use of axonal transport for studies of neuronal connectivity. Elsevier, Amsterdam, pp 114–125

80. Schwab ME, Javoy-Agid F, Agid Y (1978) Labeled wheat germ agglutinin (WGA) as a new, highly sensitive retrograde tracer in the rat brain hippocampal system. Brain Res 152: 145–150

81. Shiosaka S, Tohyama M, Takagi H, Takahashi Y, Saitoh Y, Sakumoto T, Makagawa H, Shimizu N (1980) Ascending and descending components of the medial forebrain bundle in the rat as demonstrated by the HRP blue reaction. Exp Brain Res 39:377–388

82. Somogyi P, Hodgson AJ, Smith AD (1979) An approach to tracing neuron networks in the cerebral cortex and basal ganglia. Combination of Golgi staining, retrograde transport of HRP and anterograde degeneration of synaptic boutons in the same material. Neuroscience 4:1805–1852

82a. Spencer HJ, Lynch G, Jones RK (1978) The use of somatofugal transport of HRP for tract tracing and cell labeling. In: Robertson RT (ed) Neuroanatomical research techniques. Academic Press, London New York, pp 291–316

83. Streit P, Reubi JC (1977) A new and sensitive staining method for axonally transported HRP in the pigeon visual system. Brain Res 126:530–537

84. Sturmer C, Bielenberg K, Spatz WB (1981) Electron microscopical identification of 3,3′,5,5′-tetramethylbenzidine reacted HRP after retrograde axoplasmic transport. Neurosci Lett 23:1–5

85. Teichberg S, Bloom D (1976) Uptake and fate of HRP in axons and terminals of sympathetic neurons. J Cell Biol 70:285a

86. Turner PT, Harris AB (1974) Ultrastructure of exogenous peroxidase in cerebral cortex. Brain Res 74:305–326

87. Tweedle CD (1978) Single-cell staining techniques. In: Robertson RT (ed) Neuroanatomical research techniques. Academic Press, London New York, pp 142–174

88. Vanegas H, Holländer H, Distel H (1978) Early stages of uptake and transport of HRP by cortical structures, and its use for the study of local neurons and their processes. J Comp Neurol 177:193–212

89. Walberg F, Nordby T, Dietrichs E (1980) A note on the anterograde transport of HRP within olivocerebellar fibres. Exp Brain Res 40:233–236

90. Watkins LR, Griffin G, Leichnetz GR, Mayer DJ (1980) The somatotopic organization of the *nucleus raphe magnus* and surrounding brain stem structures as revealed by HRP slow release gels. Brain Res 181:1–15

91. West JR, Black AC (1979) Enhancing the anterograde movement of HRP to label sparse
    neuronal projections. Neurosci Lett 12:35—40
92. Westrum LE, Canfield RC, O'Connor TA (1980) Projections from dental structures to the
    brain stem trigeminal complex as shown by transganglionic transport of HRP. Neurosci Lett
    20—31—36
93. Winer JA (1977) A review of the status of the HRP method in neuroanatomy. Biobehav Rev
    1:45—54

Note Added in Proof:

During the last months the following books appeared, in which HRP techniques are reviewed:
Heym Ch, Forssman W-G (eds) (1981) Techniques in neuroanatomical research. Springer, Berlin
    Heidelberg New York
Heimer L, RoBards M (eds) (1982) Neuroanatomical tract-tracing methods. Plenum Press,
    New York London
Mesulam M-M (ed) (1982) Tracing neural connections with horseradish peroxidase. Wiley & Sons,
    Sussex

Axonal Transport of Fluorescent Compounds in the Brain and Spinal Cord of Cat and Rat

ANDREAS ASCHOFF [1], NORBERT FRITZ [2], and MICHAEL ILLERT [2]

It is well known that small molecules like amino acids and nucleotides may use axonal transport [16, 20, 34, 38]. The same applies for fluorescent compounds which have a molecular weight between 400 and 1000. Whereas it was assumed in earlier studies that it might be favourable to couple them to proteins as vehicles [8, 23], it turned out later that they are more efficiently transported when used alone [12, 27].

Recently the axonal transport of fluorescent compounds became a tool in tracing neuronal connections in the central [4, 6, 26, 27] and peripheral nervous system [4, 6, 22], especially when trying to delineate collateral projections of one and the same neuron [7, 9, 21, 22, 37]. In order to evaluate the specific behaviour of the fluorescent dyes and to compare them with a well established tracer like horseradish peroxidase (HRP), we have performed a systematic analysis of dyes in the visual system of the rat and in the spinal cord of the cat [2, 17]. It turned out that, when certain precautions are observed, some of the compounds might be valuable in neuro-anatomical research. The present study focusses on points which are of interest when judging the axonal transport of fluorescent compounds, such as diffusion of the substances out of labeled neurons, transport velocity, and direction and interaction between two fluorescent compounds at the site of uptake.

Material and Methods

In the methodological approach we followed the procedures described by Aschoff and Holländer [2] and Illert et al. [17]. We investigated compounds which were found by Kuypers et al. [26, 27] and Bentivoglio et al. [3] to be suitable for retrograde tracing (see Table 1 for list and abbreviations). They were all tested in the central visual system (CVS) of the rat and all but EB in the peripheral motor system (PMS) of the cat. In the rat 50 nl of each solution or suspension was injected into the visual cortex

1 Max-Planck-Institut für Psychiatrie, Kraepelinstr. 2, D-8000 München 40, Fed. Rep. of Germany
2 Department of Physiology, Universität München, Pettenkoferstr. 12, D-8000 München 2, Fed. Rep. of Germany

Axoplasmic Transport in Physiology and Pathology
(ed. by D.G. Weiss and A. Gorio)
© Springer-Verlag Berlin Heidelberg 1982

Table 1. List of fluorescent compounds investigated

Dye	Abbreviation	PMS [a]	CVS [b]
Nuclear yellow	NY	2.0%	1.0%
Bisbenzimide	BB	10.0%	10.0%
4',6-Diamidino-2-phenylindol-2HCl	DAPI	2.0%	2.5%
Fast blue	FB	10.0% *	5.0%
Granular blue	GB	5.0% *	5.0%
True blue	TB	5.0% *	10.0%
Primuline	PR	10.0%	10.0%
Propidium iodide	PI	3.5%	3.0%
Evans blue	EB	–	10.0%

The concentrations used are indicated in weight/volume. The substances were dissolved in distilled water or in ethylene glycol (marked by *)

[a] PMS: peripheral motor system (cat); [b] CVS: central visual system (rat)

(area 17) by means of micro-pipettes. In the cat the solution was either injected into a forelimb muscle (0.5 ml) or directly applied to transected forelimb nerves. For investigation of anterograde transport of BB, 0.1 ml of a 10% solution was injected into the vitrous body of the cat eye. After survival times of 7, 9, or 24 h (rats) and 1.5 to 4 days (rats and cats) the animals were deeply anaesthetized and transcardially perfused with 4% (rat) or 10% (cat) formalin. Sections were prepared and investigated according to the methods described by Aschoff and Holländer [2].

Results

Following application of the fluorescent compounds, retrogradely labeled neurons were observed in both the PMS of the cat and the CVS of the rat. In the rat they were located in the thalamus and in the contralateral cortex ([2], see also [33]); in the cat in the respective motornuclei [11, 17]. Each compound was characterized by a distinct intracellular location of the fluorescence [2, 17] which was in most cases in agreement with the observations reported in [3, 4, 26, 27]. With the exception of BB and NY, the location of the different tracers in the cell bodies was basically identical in the CVS and PMS. The fluorescence was mainly located in the cytoplasm of the neurons. Predominantly it was bound to large granules, to a minor extent it was also diffusely distributed. PI and FB regularly labeled the nucleolus. A weak homogeneous fluorescence of the nucleus was observed with FB and TB. Interesting differences between both systems were observed with BB and NY, which are both regarded as typical nuclear stains [4, 7, 29]. In the CVS there was no doubt about the nuclear location of these dyes with a faint cytoplasmic fluorescence only. In the PMS, however, the fluorescence was present in large cytoplasmic granules and it remained doubtful if the nucleus was labeled (an intense fluorescence of the nucleolus was always observed). This different distribution of the label was independent of the pH of the fixation medium (media of pH 4.0, 6.6, 7.2, and 8.0 were tested) and of the survival time (CVS: 9 h—4 days; PMS: 1.5—7 days). It was also invariable with different BB concentrations in the injection fluid (cats, 0.25—10%). Apparently there is a difference in central neurons and spinal cord motoneurons regarding the intracellular distribution of BB and NY after their arrival to the cell soma.

In the PMS the velocity of the retrograde transport was calculated only approximately. Labeled motoneurons were observed after a survival time of three days. With a nerve length between 15 and 20 cm, this would correspond to a velocity of at least 60 mm/day. In two experiments with intramuscular injection of BB we used a survival time of only 36 h and observed in both cases tracer positive neurons in the parent motornucleus. This corresponds to a transport velocity of more than 100 mm/day for BB, and it remains to be investigated whether that of the other substances is in a comparable range. Corresponding values were reported by Glatt and Honegger [12] for the triceps nerve of the rat (70 mm/day, EB/albumine complex) and by Kristensson et al. [24] for the hypoglossus nerve of the rabbit (120 mm/day, EB-albumine complex). A more detailed analysis has been performed in the CVS, where the time course of retrograde labeling with BB and NY was estimated in the dorsal part of the lateral geniculate nucleus (LGNd). The first tracer positive neurons were seen after a survival time of 7 or 9 h, which corresponds to a transport velocity of about 60 mm/day (length of geniculo-cortical projection about 20 mm). After longer survival periods the area labeled with BB or NY showed different spatial extensions (for a quantitative analysis the area covered by labeled neurons was drawn on paper and its size measured with a planimeter (see [2]). Both tracers labeled an area of 0.5 mm^2 when the animals survived dye injection for 9 h. After longer survival times the area covered by NY positive neurons did not change significantly, whereas that covered by BB positive neurons increased (Fig. 1). It was maximal (2 mm^2) after 24 h, but decreased when longer survival times were used. A comparable analysis performed with HRP [14] shows striking correspondence with these data.

It was found in the CVS of the rat that BB and NY are anterogradely transported [2]. This was indicated by the presence of tracer positive nuclei of nerve and glial cells in the ventral part of the LGN and in the *superior culliculus*. Since neither of these structures project to the visual cortex [15, 31, 33, 36] it was proposed that anterogradely transported BB and NY had leaked out of the terminals into the surrounding tissue and stained neuronal and glial nuclei. This assumption was tested in the cat retino-geniculate projection. Injection of BB into the eye labeled neuronal and glial nuclei in lamina A1 of the ipsilateral LGNd and in lamina A and A1 of the contralateral LGNd (Fig. 2). The presence of tracer positive neurons very strongly supports the hypothesis that BB is anterogradely transported into nerve terminals and secondarily moves to neighbouring neurons and glial cells. Other fluorescent compounds were not tested in this preparation, but from the results obtained in the rat CVS [2], it is likely that NY is also anterogradely transported. The situation might be different with regard to the diamidino-compounds (DAPI, FB, GB, TB) and PI, since in the thalamus or *superior colliculus* no signs were found which would indicate their anterograde transport.

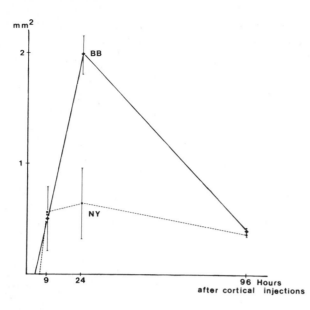

Fig. 1. Quantity of cells in the dorso-lateral thalamus of the rat labeled with BB or NY after different survival times. We estimated that 0.5 mm² was covered by about 400 labeled neurons. The size of the area with labeled cells in the different experiments (n) is experessed as means ± S.D. For 9 h survival times n were 5 (BB) and 3 (NY), for 24 h survival times n were 4 (BB) and 3 (NY) and for 96 h survival times n were 5 (BB) and 3 (NY)

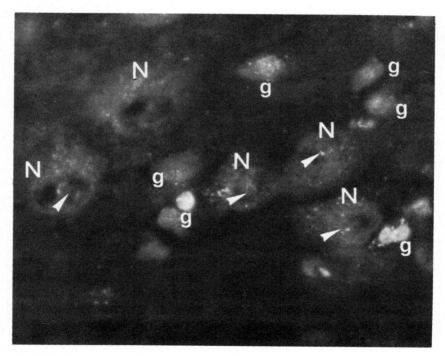

Fig. 2. BB labeled neuronal *(N)* and glial *(g)* nuclei in lamina A of the contralateral LGNd of the cat after injection into the eye. In most neuronal nuclei a fluorescent nucleolar ring *(arrows)* can be seen

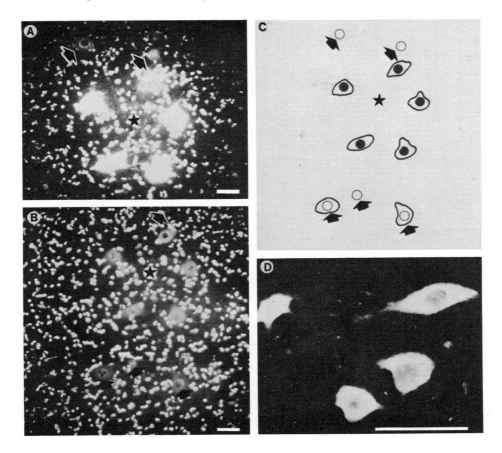

Fig. 3. Transneuronal movements of BB. A Cross-section through a motor nucleus with a cluster of 5 neurons labeled by retrogradely transported BB (the center of the cluster is marked by *) and two neurons *(arrows)* labeled by diffusion of BB out of retrogradely labeled neurons. This section was investigated immediately after mounting in aqueous medium. B Same section as shown in A with the cluster of primarily labeled neurons (*) but examined 24 h later. The three secondarily labeled neurons *(arrows)* in the lower part of the picture were not labeled immediately after mounting (not shown in A). C Schematic drawing of the section shown in A and B. Primarily labeled neurons (•), secondarily labeled neurons (○), *arrows* as in A and B. D A section from the same material but dry mounted and photographed several days after mounting. Note the sparse glial labeling. *Bars* 100 μm

Transneuronal movement of BB might be due to its passive diffusion into the extracellular space. We have in fact observed a high diffusibility of this compound and of the chemically related NY. This was easily demonstrated in sections which had been mounted in an aqueous medium. Fig. 3A shows a cross section through a motornucleus labeled with BB (the photograph has been taken immediately after mounting). There is a group of five intensely fluorescent motoneurons. In addition there are strongly fluorescent nuclei of glial cells which are located within the neuronal cell cluster and in the region immediately surrounding it. In these sections we regularly observed neurons which had a rather weak fluorescence and were located in the neighbourhood of a strongly fluorescent cell cluster (arrows). Motoneurons fluorescent at such a low intensity were practically never observed when the sections were treated according to the procedure described by Aschoff and Holländer [2] (a section from that material is illustrated in Fig. 3D). We therefore assume that tracer diffusion from the intensely labeled motoneurons of Fig. 3A has produced a secondary labeling of the glial nuclei and of the weakly fluorescent motoneurons. Fig. 3B shows the same section as in A but 24 h later. The area with labeled glial nuclei and secondarily labeled neuronal nuclei is markedly enlarged. In some cells there is also a faint, homogeneous fluorescence of the cytoplasm. In addition, the intensity of the primary labeled motoneurons has drastically decreased.

Table 2. A Labeling of ECU motoneurons following intramuscular injection[ECU(M)] of a solution consisting of BB and PI (two experiments. The numbers (means ± S.D.) give the tracer positive neurons in the ECU motornucleus. 402 neurons were double-labeled with both substances (Σ), 11 were single labeled with BB, 8 single labeled with PI. HRP (three experiments) labeled a mean of 432 cells.
B Labeling of ECU and EDC motoneurons with HRP, PI, and PI in combination with BB (different experiments for each substance). ECU motoneurons were labeled by intramuscular injection [ECU(M)], EDC motoneurons by tracer application to the transected nerve [EDC(N)]. The number of tracer positive neurons in the different experiments (n) is expressed in means ± S.D. Note that PI labels more neurons in the respective nucleus when applied together with BB, than when PI is applied alone.
C Labeling of EDC motoneurons with BB following application of the tracer to the transected nerve [EDC(N)]. BB, when applied alone to the EDC(N), did not label any motoneurons. In combination with PI and DMSO a fraction of the nucleus was labeled. The same was the case when HRP was added to the injection solution but quantative data were not obtained from these experiments. The number of cells labeled with HRP using the same mode of application is given for comparison

		Σ	BB	PI	HRP	
A	ECU(M)	402 ± 16 n = 2	11 ± 11	8 ± 1	432 ± 41 n = 3	

		HRP	PI	PI(+BB)	
B	ECU(M)	432 ± 41 n = 3	308 ± 28 n = 5	392 ± 4 n = 2	
	EDC(N)	405 ± 68 n = 3	264 ± 42 n = 4	353 ± 14 n = 2	

		HRP	BB	BB(+PI)	BB(+DMSO)	BB(+HRP)
C	EDC(N)	405 ± 68 n = 3	O n = 4	105 ± 21 n = 2	170 n = 1	+

One of the main applications of the fluorescent dyes is their use in experiments which aim to establish collateral projections of one and the same neuron [3, 9, 21, 22, 26]. For this purpose two or three compounds which are fluorescent at easily distinguishable wave lengths are applied to different suspected terminal regions of the system in question. After appropriate survival times the somata of the neurons are investigated with respect to the presence of the different compounds in one cell. To test for a possible interaction between the various substances we have used BB and PI in the PMS by applying them at the same time to a common site. This has led to double labeling of nearly all motoneurons composing the respective motornucleus. Table 2A summarizes the data obtained with injection of the substances in the *extensor carpi ulnaris* muscle (ECU). With HRP injections about 430 neurons were traced in that nucleus [11, 17]. When a solution consisting of BB and PI was applied, about 400 of these neurons were double labeled from both substances, 11 neurons single labeled with BB, 8 neurons single labeled with PI. With that approach we also observed interactions between both fluorescent compounds. Thus the number of neurons labeled with PI was higher when this substance was applied in the presence of BB than when it was applied alone. Table 2B compares the cell numbers obtained with HRP, PI, and PI in the presence of BB for muscle and nerve application. The higher number of PI positive cells which is observed in the presence of BB is probably due to the fact that BB acts as a solvent. With a concentration of 3.5% in distilled water, PI is a suspension which becomes clear when BB is added (no sediment is visible after centrifugation). This could imply that PI, when applied in combination with BB, is present at the axonal uptake sites in higher concentration. To test this assumption we have centrifuged a suspension consisting of 3.5% PI and added BB (final concentration 10%) to the supernatant. When this solution was injected into the ECU or applied to a transected nerve the number of PI positive cells was comparable to that obtained with PI alone.

We also observed an action of PI on BB. When BB was injected into the ECU the number of BB labeled motoneurons was independent of the presence of PI in the injection fluid. The situation was different in case of BB application to a transected nerve (Table 2C). In that application mode BB (as well as NY and PR) does not label any motoneurons [17]. However, when used in combination with PI, 105 BB positive cells were counted, which is about 25% of the neurons traceable with HRP (the number of PI positive cells was in the range shown in Table 2B). They were scattered throughout the whole nucleus and there was no preference for a particular position. Both large and smal neurons were observed. They were all double labeled with PI, neurons single labeled with BB were not observed. PI was not the only substance which led to a visualization of BB when applied together with this compound to a transected nerve. Similar results were observed with DMSO and HRP (Table 2C).

Discussion

All the investigated fluorescent compounds were principally taken up by the neurons and were axonally transported. This is in agreement with the numerous reports in the literature (e.g. [4, 5, 27]). There is apparently no difference between different neuronal systems since fluorescent compounds have been reported to label a variety of central systems (e.g. visual system: this study and [2, 18]; cortico-cortical connections: [7]; basal ganglia: [21, 22, 28]; hypothalamic neurosecretory system: [37], spinal cord motoneurons (this study and [10, 17, 28]) and spinal preganglionic sympathetic neurons [9]. This uniform behaviour points to an unspecific uptake mechanism similar to the one discussed for HRP and other macromolecular substances [13, 35]. There is one notable exception, that is the absence of retrograde labeling of motoneurons following BB, NY and PR application to a transected nerve. This finding seems to be specific for motoneurons, since BB application to a transected *ramus communicans albus* labels preganglionic sympathetic neurons in their usual location and in normal quantitative amounts [9]. The differential effects observed with BB, NY and PR after intramuscular injection, motor nerve or preganglionic nerve application could be due to the different myelin content of the respective preparations which is the most conspicuous difference between them (intramucular injections: unmyelinated terminal membranes; transected preganglionic nerves: poorly myelinated B-fibres; motor nerves: heavily myelinated Aα- and Aγ-fibres). The implications of these findings for the use of the respective substances in neuroanatomical research are discussed in detail in Illert et al. [17].

After application to a transected motor nerve BB was transported only when other substances were added to the injection solution. Since this compound is principally transported in motor axons this facilitatory effect should occur at the uptake site. There are many possible explanations. It is known for example that large basic molecules facilitate the uptake of small ones [32]. Such effects have been described for poly-L-orthithine (e.g. facilitation of inuline uptake [1]) and in fact this substance is used in neuroanatomical research for this purpose [14]. Possibly a similar mechanism applies for the HRP-BB interaction. The effect of PI on BB is more difficult to understand since both substances have a comparable molecular weight. The labeling of all BB positive motoneurons with PI could indicate a specific interaction between both substances at the axonal membrane. The action of DMSO is probably a rather unspecific effect. This substance is known to change the permeability of membranes [19, 30] which subsequently may lead to an enhanced BB transfer into the axons.

All compounds were retrogradely transported. Regarding the transport velocity the methods in use were certainly not discriminative enough to give precise data, but they indicate that the tracers move in the fast transport compartment [13]. In this respect they seem to be very similar to HRP. A contribution of a slow transport compartment to the labeling of the neurons cannot be ascertained from our results. There is the interesting observation, however, that the fluorescent intensity of neurons labeled with diamidino-compounds increases with time [2, 5] which should be expected if they move in a slow transport compartment in addition to a fast one.

In agreement with data available from the literature [2, 4, 26] we found evidence for anterograde transport of BB and NY. However, this was not possible by the direct demonstration of these substances in the axon terminals, but only by their presence in nuclei of surrounding neurons and glial cells in regions which only received afferent fibres from the site of injection. Similar observations were not made with other fluorescent compounds.

In spinal motoneurons the intrasomatic location of the different compounds was rather uniform. The substances were mainly deposited in large granules in the cytoplasm, some of them were also found in the nucleolus (BB, NY, FB, PI). The cytoplasmic location is in agreement with the distribution of the compounds (FB, GB, TB, BB, PI) in central neurons as observed in freeze-dried preparations [2, 6]. Thus it seems that the primary location of the fluorescent compounds is cytoplasmic. With the exception of BB and NY the intrasomatic location of the compounds was basically similar in the CVS and PMS. BB and NY, however, with their prominent nuclear location in the CVS had a completely different distribution in the PMS. The labeling of the nucleus (and probably also of the nucleolus) in the CVS seems to be a secondary phenomenon due to diffusion of the tracers from their primary cytoplasmic deposit. This is supported by the finding that a nuclear BB and NY fluorescence is achieved in spinal motoneurons when the in vitro diffusion is provoked by embedding the sections in an aqueous medium. It is, however, difficult to explain why this secondary diffusion is prominent in central neurons and practically nonexistent in motoneurons when identical mounting procedures are used. The findings indicate that the intracellular location of BB and NY is not stable and may be different in different neuronal systems. This has obvious implications for the use of these substances in neuroanatomical research [2, 17, 25].

Conclusions

In their axonal transport the fluorescent compounds show many similarities with HRP. They are retrogradely transported as well as anterogradely, they are mainly localized in the cytoplasm with a granular appearance and most of them have, within certain limits, a comparable sensitivity [17]. In some respects, such as diffusion and uptake, there seem to be differences which may be explained by the different molecular weights and different chemical composition of these substances. Thus it appears that the fluorescent compounds and HRP use a common transport system in which they become incorporated via nonspecific uptake.

Acknowledgments. We gratefully acknowledge the support and advice given by Prof. H. Holländer during the course of this investigation. Miss Lucia Schindler and Mr. R. Wörndl provided excellent technical assistance. The study was supported by the Deutsche Forschungsgemeinschaft (Il 13/8-1).

References

1. Amundsen E, Ryser HJ-P (1968) Unpublished observations. In: Ryser HJ-P: The uptake of foreign proteins by mammalian cells and the functions of pinocytosis. Bull Schweiz Akad Med Wiss 24:363–385
2. Aschoff A, Holländer H (1982) Fluorescent compounds as retrograde tracers compared with horseradish peroxidase (HRP). I. A parametric study in the central-visual system of the albino rat. J Neurosci Methods (in press)

3. Bentivoglio M, Kooy D van der, Kuypers HGJM (1979) The organization of the efferent projections of the substantia nigra in the rat. A retrograde fluorescent double labeling study. Brain Res 174:1–17

4. Bentivoglio M, Kuypers HGJM, Catsman-Berrevoets CE (1980) Retrograde neuronal labeling by means of bisbenzimide and nuclear yellow (Hoechst S 769121). Measures to prevent diffusion of the tracers out of retrogradely labeled neurons. Neurosci Lett 18:19–24

5. Bentivoglio M, Kuypers HGJM, Catsman-Berrevoets CE, Dann O (1979) Fluorescent retrograde neuronal labeling in rat by means of substances binding specifically to adenine-thymine rich DNA. Neurosci Lett 12:235–240

6. Björklund A, Skagerberg G (1979) Simultaneous use of retrograde fluorescent tracers and fluorescence histochemistry for convenient and precise mapping of monoaminergic projections and collateral arrangements in the CNS. J Neurosci Methods 1:261–277

7. Catsman-Berrevoets CE, Lemon RN, Verburgh CA, Bentivoglio M, Kuypers HGJM (1980) Absence of callosal collaterals derived from rat corticospinal neurons. A study using fluorescent retrograde tracing and electrophysiological techniques. Exp Brain Res 39:433–440

8. Clasen RA, Pandolfi S, Hass GM (1970) Vital staining, serum albumin and the blood-brain barrier. J Neuropathol Exp Neurol 29:266–284

9. Deuschl G, Illert M, Aschoff A, Holländer H (1981) Single preganglionic sympathetic neurons of the cat branch intraspinally and project through different *rami communicantes albi* – a retrograde double labelling study with fluorescent tracers. Neurosci Lett 21:1–5

10. Enerbäck L, Kristensson K, Olsson T (1980) Cytophotometric quantification of retrograde axonal transport of a fluorescent tracer (primuline) in mouse facial neurons. Brain Res 186:21–32

11. Fritz N, Illert M, Saggau P (1981) Location of dorsal interosseus motor nuclei in the cat. Neurosci Lett 21:243–248

12. Glatt HR, Honegger CG (1973) Retrograde axonal transport for cartography of neurones. Experientia 29:1515–1517

13. Grafstein B, Forman DS (1980) Intracellular transport in neurons. Physiol Rev 60:1167–1283

14. Hadley RT, Trachtenberg MC (1978) Poly-1-ornithine enhances the uptake of horseradish peroxidase. Brain Res 158:1–14

15. Hughes HC (1977) Anatomical and neurobehavioral investigations concerning the thalamo-cortical organization of the rat's visual system. J Comp Neurol 175:311–336

16. Hunt SP, Künzle H (1976) Bidirectional movement of label and transneuronal transport phenomena after injection of ^{3}H-adenosine into the central nervous system. Brain Res 112:127–132

17. Illert M, Fritz N, Aschoff A, Holländer H (1982) Fluorescent compounds as retrograde tracers compared with horseradish peroxidase (HRP). II. A parametric study in the peripheral motor system of the rat. J Neurosci Methods (in press)

18. Illing R-B (1980) Axonal bifurcation of cat retinal ganglion cells as demonstrated by retrograde double labelling with fluorescent dyes. Neurosci Lett 19:125–130

19. Keefer DA (1978) Horseradish peroxidase as a retrogradely-transported, detailed dendritic marker. Brain Res 140:15–32

20. Kerkut GA, Shapira A, Walker RJ (1967) The transport of ^{14}C-labelled material from CNS ⇌ muscle along a nerve trunk. J Comp Biochem Physiol 23:729–748

21. Kooy D van der (1979) The organization of the thalamic, nigral and raphe cells projecting to the medial vs lateral caudate-putamen in rat. A fluorescent retrograde double labeling study. Brain Res 169:381–387

22. Kooy D van der, Hattori T (1980) Dorsal raphe cells with collateral projections to the caudate-putamen and substantia nigra: a fluorescent retrograde double labeling study in the rat. Brain Res 186:1–7

23. Kristensson K (1970) Transport of fluorescent protein tracer in peripheral nerves. Acta Neuropathol 16:293–300

24. Kristensson K, Olsson Y, Sjöstrand J (1971) Axonal uptake and retrograde transport of exogenous proteins in the hypoglossal nerve. Brain Res 32:399–406

25. Kuypers HGJM, Bentivoglio M, Catsman-Berrevoets CE, Bharos AT (1980) Double retro-grade neuronal labeling through divergent axon collaterals, using two fluorescent tracers with the same excitation wavelength which label different features of the cell. Exp Brain Res 40:383–392

26. Kuypers HGJM, Bentivoglio M, Kooy D van der, Catsman-Berrevoets CE (1979) Retrograde transport of bisbenzimide and propidium iodide through axons to their parent cell bodies. Neuroci Lett 12:1–7

27. Kuypers HGJM, Catsman-Berrevoets CE, Padt RE (1977) Retrograde axonal transport of fluorescent substances in the rat's forebrain. Neurosci Lett 6:127–135

28. Kuzuhara S, Kanazawa I, Nakanishi T (1980) Topographical localization of the onuf's nuc-lear neurons innervating the rectal and vesical striated sphincter muscles: a retrograde fluor-escent double labeling in cat and dog. Neurosci Lett 16:125–130

29. Latt SA (1974) Detection of DNA synthesis in interphase nuclei by fluorescence microscopy. J Cell Biol 62:546–550

30. Martin D, Hanthal HG (1975) Dimethyl sulphoxide (translated by Halberstadt ES). Van Nostrand Reinhold, New York, pp 500

31. Ribak CE, Peters A (1975) An autoradiographic study of the projections from the lateral geniculate body of the rat. Brain Res 92:341–368

32. Ryser HJ-P (1968) The uptake of foreign proteins by mammalian cells and the functions of pinocytosis. Bull Schweiz Akad Med Wiss 24:363–385

33. Schober W, Lüth H-J, Gruschka H (1976) Die Herkunft afferenter Axone im striären Kortex der Albinoratte: Eine Studie mit Meerrettich-Peroxidase. Z Mikrosk Anat Forsch 90:399–415

34. Schubert P, Kreutzberg GW (1975) ^{3}H-adenosine, a tracer for neuronal connectivity. Brain Res 85:317–319

35. Schwab M, Thoenen H (1977) Selective trans-synaptic migration of tetanus toxin after retrograde axonal transport in peripheral sympathetic nerves: a comparison with nerve growth factor. Brain Res 122:459–474

36. Swanson LW, Cowan WM, Jones EG (1974) An autoradiographic study of the efferent con-nections of the ventral lateral geniculate nucleus in the albino rat and the cat. J Comp Neurol 156:143–164

37. Swanson LW, Kuypers HGJM (1980) The paraventricular nucleus of the hypothalamus: Cytoarchitectonic subdivisions and organization of projections to the pituitary, dorsal vagal complex, and spinal cord as demonstrated by retrograde fluorescence double-labeling methods. J Comp Neurol 194:555–570

38. Wise SP, Jones EG (1976) Transneuronal or retrograde transport of ^{3}H adenosine in the rat somatic sensory system. Brain Res 107:127–131

Subject Index

Axoplasmic Transport

edited by Dieter G. Weiss

The publication of a treatise covering basic and applied aspects of axoplasmic transport was conceived at the occasion of the Workshop on Axoplasmic Transport which was held at Schloss Elmau in Bavaria in 1981. The more basic aspects of axoplasmic transport research are covered in this volume which is published concomitantly.

Section 3. General Characterization of Axoplasmic Transport: Materials and Properties

Axoplasmic Transport

Editor: **D. G. Weiss**
1982. 181 figures. XIII, 477 pages
(Proceedings in Life Sciences)
ISBN 3-540-11662-1

This book is the first volume in which all basic aspects of axoplasmic transport and its mechanisms are comprehensively discussed by the experts. The necessity of using multi-disciplinary approaches to obtain a clear understanding is especially obvious in this field, lying as it does on the frontiers of neurochemistry, cell biology, neurophysiology and biophysics.

The book is a complete treatise on the properties of axoplasmic transport in a variety of neuronal systems, on its physical and biochemical implications, and on our secured knowledge as well as the current hypotheses about axoplasmic transport mechanisms.

During the last few years the impact of axoplasmic transport research on many fields of cell biology and neuroscience has increased dramatically. Therefore, more and more neurochemists, neuroanatomists, neuropathologists, neurophysiologists, neurologists and cell biologists need to have an understanding of the present status of axoplasmic transport. It is for these scientists that this book is of vital interest.

International Cell Biology 1980–1981

Papers Presented at the Second International Congress on Cell Biology, Berlin (West), August 31–September 5, 1980
Editor: **H. G. Schweiger**
1981. 595 figures. XVIII, 1033 pages
ISBN 3-540-10475-5

International Cell Biology 1980-1981 contains contributions presented at the Second International Congress on Cell Biology held in West Berlin, August 31 – September 5, 1980. The authors of the contributions were selected as speakers for the Congress for their leading role in their respective fields. The topics cover a uniquely broad range of research areas, providing an excellent reflection of the present status of cell biology. This book will remain a useful source of information to biologists, medical researchers and biochemists for years to come.

Neuronal-glial Cell Interrelationships

Editor: **T. A. Sears**
Report of the Dahlem Workshop on Neuronal-glial Cell Interrelationships: Ontogeny, Maintenance, Injury, Repair, Berlin 1980, November 30 – December 5
Rapporteurs: **R. L. Barchi, G. R. Strichartz, P. A. Walicke, H. L. Weiner**
Program Advisory Commitee: **T. A. Sears** (Chairman); **A. J. Aguayo, B. G. W. Arnason, H. J. Bauer, B. N. Fields, W. I. McDonald, J. G. Nicholls**
1982. 5 photographs, 13 figures, 8 tables. X, 375 pages
(Dahlem Workshop Reports-Life Sciences Research Report, Volume 20)
ISBN 3-540-11329-0

R. F. Schleif, P. C. Wensink

Practical Methods in Molecular Biology

1981. 49 figures. XIII, 220 pages
ISBN 3-540-90603-7

Contents: Using *E. coli*. – Bacteriophage Lambda. – Enzyme Assays. – Working with Proteins. – Working with Nucleic Acids. – Constructing and Analyzing Recombinant DNA. – Assorted Laboratory Techniques. – Appendix I: Commonly Used Recipes. – Appendix II: Useful Numbers. – Bibliography. – Index.

Springer-Verlag
Berlin
Heidelberg
New York

Journal of Comparative Physiology · A+B

Founded in 1924 as
Zeitschrift für vergleichende Physiologie
by **K. von Frisch** and **A. Kühn**

The Journal of Comparative Physiology publishes original articles in the field of animal physiology. In view of the increasing number of papers and the high degree of scientific specialization the journal is published in two sections.

A. Sensory, Neural, and Behavioral Physiology
Physiological Basis of Behavior; Sensory Physiology; Neural Physiology; Orientation, Communication; Locomotion; Hormonal Control of Behavior

B. Biochemical, Systemic, and Environmental Physiology
Comparative Aspects of Metabolism and Enzymology; Metabolic Regulation, Respiration and Gas Transport; Physiology of Body Fluids; Circulation; Temperature Relations; Muscular Physiology

Human Neurobiology

Editorial Board:
D. H. Ingvar, M. D., (Chairman of the Editorial Board), Lund; **G. Baumgartner,** M. D., Zürich; **W. E. Bunney,** M. D., Bethesda; **O. J. Grüsser,** M. D., Berlin; **B. Julesz,** M. D., Murrey Hill; **D. Kimura,** Ph. D., London, Canada; **M. Mishkin,** M. D., Bethesda; **I. Rentschler,** Dr. rer. nat., (Managing Editor), München; **M. Zimmermann,** Dr.-Ing., Heidelberg

Human Neurobiology is primarily devoted to studies of the biological basis of mental activity and human behavior. Though the major emphasis is on the species of man, it also publishes investigations on other primates related to human neurobiology. **Human Neurobiology** attempts to cover and integrate neuropsychology, sensory physiology, neuro-endocrinology, neurology, neurosurgery, neurophysiology, neuropsychiatry, neurolinguistics, neuroethology, developmental neurobiology, neurochemistry, neuropharmacology, neuroanatomy, and neuropathology.
It will thus form a bridge between basic and applied science within the fields mentioned. **Human Neurobiology** publishes issues containing independent original articles as well as reviews. Some issues are devoted to special topics.

Experimental Brain Research

Coordinating Editor: O. Creutzfeldt, Göttingen

Research on the central nervous system has developed during the last 20 years into a broad interdisciplinary field. Experimental Brain Research was one of the first journals to represent this interdisciplinary approach, and it is still leading in this field. The journal attracts authors from all over the world and is distributed internationally. It covers the whole field of experimental brain research and thus reflects a variety of interests as represented by the International Brain Research Organization (IBRO): neurobiology of the brain.
Experimental Brain Research accepts original contributions on any aspect of experimental brain research relevant to general problems of cerebral function in the fields of: neuroanatomy, sensory- and neurophysiology, neurochemistry, neuropharmacology, developmental neurobiology, experimental neuropathology, and behavior.
Short Research Notes permit speedy publication of interesting findings of on-going research. Short abstracts of meetings are published in running issues, and more extensive meeting reports appear in supplementary volumes of Experimental Brain Research.

Subscription information and/or **sample copies** are available from your bookseller or directly from Springer-Verlag, Journal Promotion Dept., P. O. Box 105 280, D-6900 Heidelberg, FRG

Springer-Verlag
Berlin
Heidelberg
New York